FINITE ELEMENT
MESH GENERATION

Saxe-Coburg Publications on Computational Engineering

Computational Modelling of Masonry, Brickwork and Blockwork Structures
Edited by: J.W. Bull

Innovative Computational Methods for Structural Mechanics
Edited by: M. Papadrakakis and B.H.V. Topping

Parallel and Distributed Processing for Computational Mechanics: Systems and Tools
Edited by: B.H.V. Topping

High Performance Computing for Computational Mechanics
Edited by: B.H.V. Topping and L. Lämmer

Computational Mechanics using High Performance Computing
Edited by: B.H.V. Topping

Object Oriented Methods and Finite Element Analysis
R.I. Mackie

Computational Mechanics for the Twenty-First Century
Edited by: B.H.V. Topping

Derivational Analogy Based Structural Design
B. Kumar and B. Raphael

Parallel Finite Element Computations
B.H.V. Topping and A.I. Khan

FINITE ELEMENT MESH GENERATION

B.H.V. Topping, J. Muylle, P. Iványi
R. Putanowicz and B. Cheng

SAXE-COBURG
PUBLICATIONS

published 2004 by
Saxe-Coburg Publications
Civil-Comp Ltd, Dun Eaglais, Station Brae
Kippen, Stirling, FK8 3DY, Scotland

Saxe-Coburg Publications is an imprint of Civil-Comp Ltd

ISBN 1-874672-10-5

British Library Cataloguing in Publication Data
A catalogue record for this book is available from the British Library

The paving algorithm, described in Chapter 5, is protected by US patent 5,315,537, May 1994. The rights of Sandia National Laboratories, Albuquerque, New Mexico, USA and their permission to describe this algorithm are acknowledged.

Printed in Great Britain by Bell & Bain Ltd, Glasgow

Contents

Preface

This book was a collaborative project undertaken by former members of the Structural Engineering Computational Technology (SECT) Research Group at Heriot-Watt University. This group was established in 1989 in the Civil Engineering Department of Heriot-Watt University by Professor Barry Topping and Dr Asad Khan. From 1990 a major focus of this group was the MSc in Structural Engineering Computational Technology (SECT) that provided a course in state-of-the-art computational methods using the latest computational technology both hardware and software. During the years 1990 to 1998 many students completed this course and continued their interest in computational research by undertaking research for a higher degree.

The research developed by the group was greatly supported and enriched by the contribution of the MSc students. The External Examiners Professor Peter Bettes and the late Professor Ernest Hinton encouraged, supported and enriched both the MSc and our research. In the early 1990s, it became apparent that little finite element research could be undertaken without the aid of robust mesh generation software. I remember, in particular, the encouragement of Ernie Hinton, who was developing a similar mesh generation code at Swansea University within his ADaptivity and OPTimisation Research Group (ADOPT). Ernie was always very generous with his time, encouragement and his knowledge. Without him this book would not have been written. In 1995 the SECT Research Group moved to the Department of Mechanical and Chemical Engineering at Heriot-Watt University.

Many members of the SECT research group contributed to the development of the mesh generation software and associated software libraries. Some of the earliest work was undertaken by Dr Asad Khan, Dr Charles Augarde and Clare Greehan. More recently Dr János Sziveri and Colin Seale made a significant contribution by the development of the "E_Library" described in Appendix D. The contributions of Dr Janet Wilson and Dr Graham Seed are also gratefully acknowledged.

In recent years, the authors of this book enjoyed a healthy cooperation on many research projects. Although the original motivation for the development of this software was to support the MSc course, many other researchers have expressed interest in the development of these mesh generation codes. The computer codes are currently available for research use from the authors' website (www.sect.mce.hw.ac.uk).

This book is not intended to be definitive, complete or always to provide the best solution. It's a record of what we did and how we solved the problems we encountered. We have been influenced and inspired by the work of Professor Nigel Weatherill and his co-workers at Swansea University. His conference presentations and lectures are, of course, a seminal influence on all who attempt to write mesh and grid generation software. Many of his references are cited in this book.

Finally we should like to thank those at Civil-Comp Ltd and Computational Technology Solutions who helped in the realisation of this project. In particular, we would like to thank Rosemary Brodie and Judy Tait.

Barry H.V. Topping
School of Engineering
Heriot-Watt University
Edinburgh

Chapter 1

Introduction

1.1 The finite element method and mesh generation

The history of the development of finite element methods dates back to the 1950s. One of the earliest papers is the 1956 paper by Turner *et al* [28]. By the late 1960s the first books concerned with finite element methods started to appear. At this time the available memory of computers was very limited and so the size of problems that could be solved was limited hence the "design" of a mesh was not the significant task that it is today. In the early books by Zienkiewicz [30], Desai and Abel [7] and Bathe [4] there was little emphasis on mesh generation. In general, the objective was to limit the mesh size so that the finite element problem could be analysed using the then available computers that were short of memory and processing speed.

With the advent of the frontal solution [10], in 1970, the size of the problem that could be solved increased significantly. The interest in automating the mesh design process became a more significant goal. Books on the finite element method began to include some consideration of the mesh generation problem [1,5,9].

By the middle of the 1980s the memory available on main frame computers had significantly increased but the memory available with microcomputers was limited. Thus for a short time frontal solutions became attractive again but this time they were used on microcomputers.

In the early 1990s mesh generation became a significant pre-processing task for the finite element method. Research into mesh generation became a significant theme in its own right and books concerned with mesh generation were published. Automating the finite element analysis process using error estimates and adaptivity started to impact on industry in the late 1990s.

Finally, the use of parallel and distributed computing may be utilised for mesh generation as will be briefly discussed in section 1.4.

1.2 Mesh generation methods

Today, the field of finite element mesh generation encompasses numerous techniques. It is not the authors' intention to provide an all inclusive reference work of the many available methods. For this purpose publications such as references [8, 19, 20] are recommended. Instead a significant collection of key algorithms for structured two-dimensional and unstructured two- and three-dimensional mesh generation are presented.

The basic mathematical background and history of the methods are described, followed by a thorough investigation of the implementation of the methods. Meshing related subjects such as post-processing of meshes to improve the mesh quality are also discussed. In two dimensions the conversion of triangular meshes to quadrilateral meshes and *vice versa* is also a subject covered in this book. Measures for mesh quality are provided and all methods are amply illustrated with examples.

First it has to be made clear that the term *mesh generation* embodies several different tasks. The following subtasks can generally be identified: initial mesh generation, remeshing and refinement.

1.2.1 Initial mesh generation

Initial or preliminary mesh generation requires a description of the geometry of the mesh domain. The mesh generator will then create a mesh that covers the domain according to mesh size requirements specified by certain control parameters. In two dimensions the geometric description of the domain can be given as a series of coordinates of nodes plus a description of the boundary of the domain.

The boundary is specified by means of a list of nodes which when connected form the domain boundary. In three dimensions a more complicated boundary description is required.

This initial mesh is sometimes referred to as the coarse mesh since it may subsequently be improved by remeshing or refinement.

1.2.2 Remeshing

The task of remeshing is performed when a given mesh (generally referred to as the initial or coarse mesh) no longer satisfies the current criteria for the mesh size or element quality. The mesh generator will extract the boundary definition of the initial mesh. Next the interior nodes and triangles will be discarded and the mesh generation task will continue in the same way as initial mesh generation.

Alternatively rather than remeshing the complete mesh as one domain the remeshing may be undertaken on an element-by-element basis as described in Section 1.3.5.

1.2.3 Refinement

Refinement occurs when additional elements are formed within an initial mesh. The refinement may be concentrated in a distinct area or areas.

1.3 Approaches to mesh generation

The object of this section is to provide the reader with an overview of the types of finite element meshes that can be generated. In the following subsections a series of techniques will be discussed with some example meshes. In each case the reader is referred to a subsequent chapter of this book for further theory and applications.

1.3.1 Structured mesh generation

A two-dimensional structured mesh generally consists of two sets of lines. The lines in the same set do not intersect each other and they intersect the lines in the other sets only once. Figure 1.1 shows a rectangular domain that has been meshed using a uniform structured mesh consisting of 90 elements and 60 nodes. This mesh is uniform because grid size which specifies the element size is the same (i.e. uniform).

Figure 1.1: Uniform structured mesh.

By specifying a variable spacing for the grid a non-uniform or adaptive structured mesh may be generated. In Figure 1.2 an adaptive or non-uniform structured mesh of 300 elements and 176 nodes is shown.

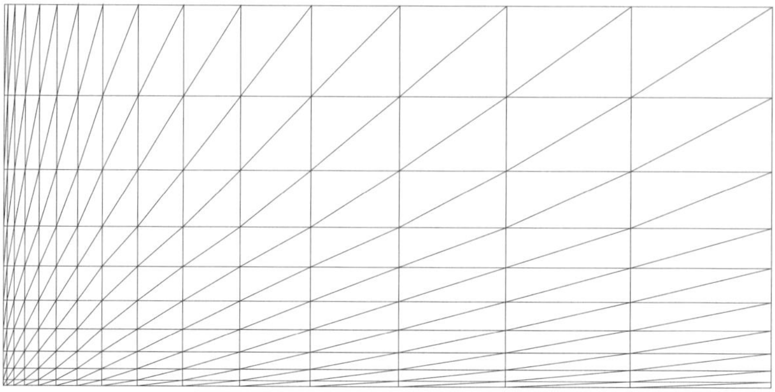

Figure 1.2: Adaptive or non-uniform structured mesh.

1.3.2 Unstructured mesh generation

In Figure 1.3 a uniform unstructured mesh of 90 elements and 60 nodes is shown. This mesh is unstructured because it is not the result of the intersection of sets of parallel lines. The mesh was generated using the advancing front technique that is described in Chapter 4.

Figure 1.3: Uniform unstructured mesh with post-processing.

The mesh shown in Figure 1.3 has been post-processed in three ways. The three ways are element triplet removal, diagonal interchange and nodal averaging. These techniques significantly enhance the quality of the mesh and will be described in Chapter 9.

1.3.3 Error analysis

The errors in a finite element analysis may arise from: mathematical modelling, finite element formulation and model discretisation. Discretisation is the process of "cutting up" the model into a number of finite elements i.e. mesh generation. The objective of error estimates is to assess how closely the finite element solution approximates the exact solution of the mathematical model [3]. In other words the error estimates assess the discretisation which is in turn related to the mesh generated for the analysis.

In the early days of finite element analysis the analyst used the results of an initial analysis in an attempt to refine the mesh, with the aim of reducing the idealisation errors. Since the late 1980s the process of assessment of these errors has been automated by the use of *a posteriori* error estimates [31].

These error estimates may be used to adaptively refine the mesh. This process was of course manual at one time but now is automated by relating the error estimate to mesh size parameter, such that the greater the error estimate the smaller the mesh size that will be generated. For a comprehensive treatment of adaptivity and refinement see Zienkiewicz and Taylor [32]. For the purposes of clarity the adaptivity referred to in the present book is referred to as h-refinement where the same class of elements are used on refinement but in some locations larger elements are formed and in other locations smaller elements are formed.

In Figure 1.4 an adaptive unstructured mesh of 188 elements and 119 nodes is shown. Again, this mesh was generated using the advancing front technique that is described in Chapter 4 and the mesh size was adaptively varied. This mesh has been post-processed in the same way as the mesh in Figure 1.3.

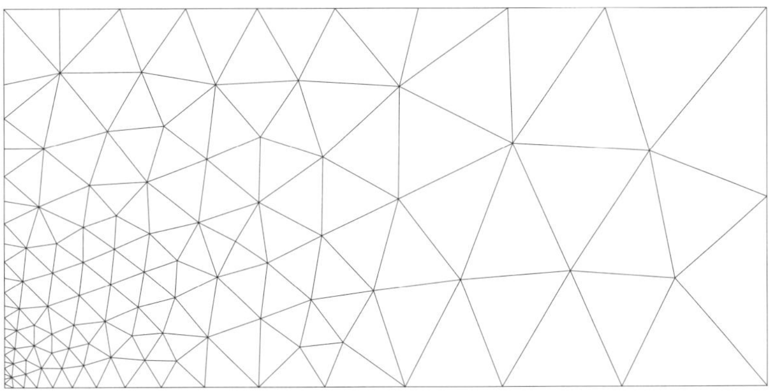

Figure 1.4: Adaptive unstructured mesh with post-processing.

1.3.4 Mesh parameters

Mesh parameters are used to specify the size of the elements to be generated within the domain. If these are specified manually approaches similar to those described in Chapter 3 may be used. Alternatively the mesh parameters may be generated from the error estimates [31]. The starting point in this book is the mesh parameters. How these are generated or calculated can be found elsewhere [2, 31].

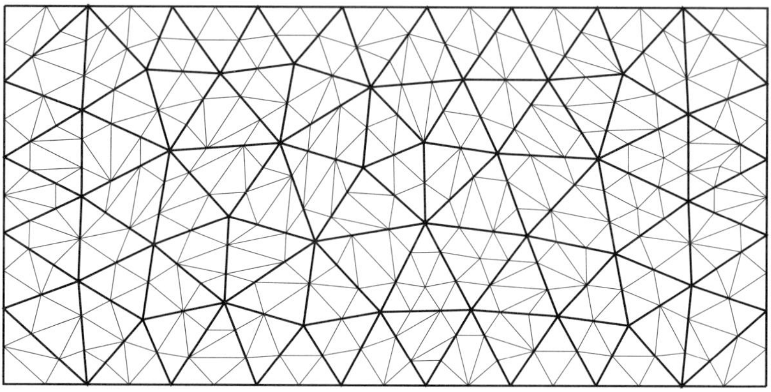

Figure 1.5: Element-by-element uniform refinement of the uniform unstructured mesh (shown in Figure 1.3) without post-processing.

1.3.5 Element-by-element refinement

The process of element-by-element refinement treats the initial coarse mesh as a series of independent meshes, each of which is equivalent to just one coarse element. Each of the coarse elements is refined separately in the same way as the initial mesh generation task. Once all coarse elements are remeshed all new triangles are merged together to form the complete refined mesh domain. As each element of the initial or coarse mesh is considered for refinement as a separate task it is not possible for an element to be enlarged to a size greater than the initial size. Thus derefinement is not possible with this strategy. Zienkiewicz and Taylor [32] refer to this type of refinement as subdivision or enrichment. With this type of refinement the original geometry of the first or coarse mesh can be seen within the refined mesh, provided mesh post-processing has not been undertaken.

The mesh shown in Figure 1.3 has subsequently been uniformly refined on an element-by-element basis using the advancing front technique. This uniform refinement is shown in Figure 1.5. This mesh has not been post-processed and consists of 506 elements and 282 nodes. The original geometry of the mesh in Figure 1.3 can be seen in the refined mesh.

Element-by-element refinement leads naturally to concepts employed in the parallelisation of mesh generation. The use of parallel and distributed processing in mesh generation will be briefly discussed in section 1.4.

1.3.6 Mesh post-processing

As an alternative the mesh shown in Figure 1.3 has been uniformly refined on an element-by-element basis using the advancing front technique. The mesh has been subsequently post-processed using element triplet removal, diagonal interchange and nodal averaging. This mesh shown in Figure 1.6, is of significantly higher quality compared to the mesh without post-processing shown in Figure 1.5. The mesh in Figure 1.6 consists of 506 elements and 282 nodes.

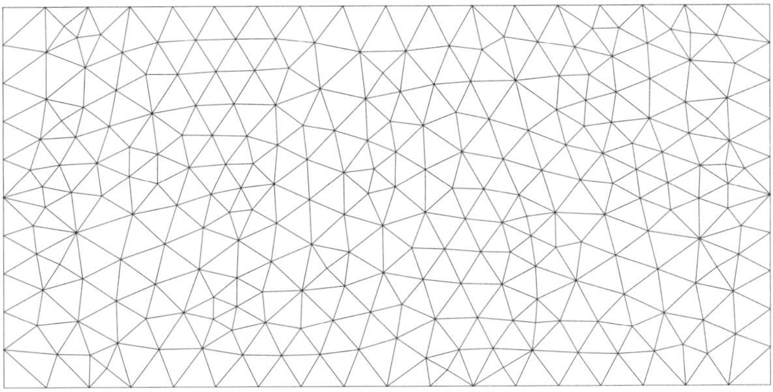

Figure 1.6: Element-by-element uniform refinement of a uniform unstructured mesh (shown in Figure 1.3) with post-processing.

The nodal averaging technique consists of taking all nodes not on the boundary of the mesh domain and placing them at the average coordinates of the nodes connected to each of them. This type of refinement is referred to as r-refinement by Zienkiewicz and Taylor [32]. Post-processing techniques are discussed in Chapter 9.

1.3.7 Adaptivity

The mesh shown in Figure 1.3 has subsequently been adaptively refined on an element-by-element basis using the advancing front technique. This adaptively refined mesh is shown in Figure 1.7. This mesh has not been post-processed and consists of 186 elements and 115 nodes.

After post-processing the mesh in Figure 1.7, is improved and the final mesh is shown in Figure 1.8.

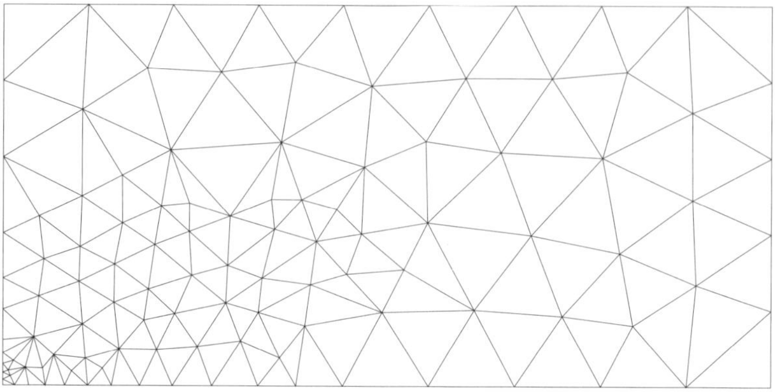

Figure 1.7: Element-by-element adaptive refinement of an uniform unstructured mesh, shown in Figure 1.3, without post-processing.

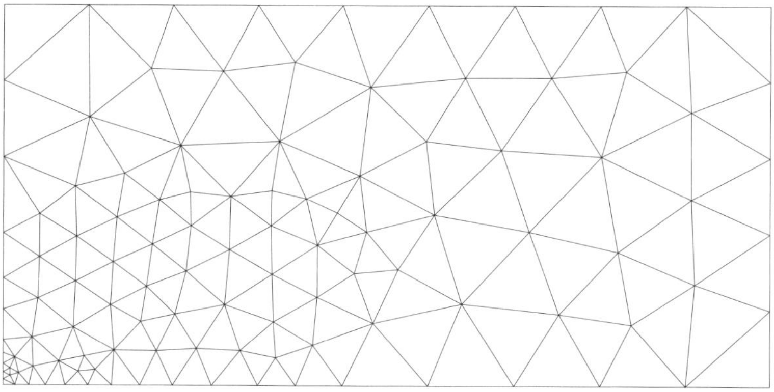

Figure 1.8: Element-by-element adaptive refinement of a uniform unstructured mesh, shown in Figure 1.3, with post-processing.

1.3.8 Quadrilateral mesh conversion and paving

All the meshes shown so far have been made up of triangles. Obviously both triangles and quadrilateral finite elements may be used. Quadrilateral meshes may be generated by converting a triangular mesh, as shown in Figure 1.4, to a quadrilateral mesh as shown in Figure 1.9. The number of elements in the converted mesh is 104 and the number of nodes is 129. Conversion procedures will be discussed in Chapter 8.

An alternative quadrilateral mesh may be generated using the paving method described in detail in Chapter 5. This method generates quadrilaterals directly without

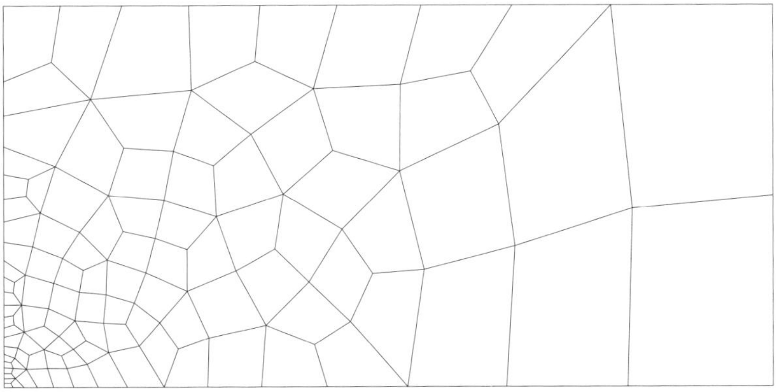

Figure 1.9: Conversion, using the fission element technique, of the adaptive unstructured triangular mesh, shown in Figure 1.4, to a quadrilateral mesh.

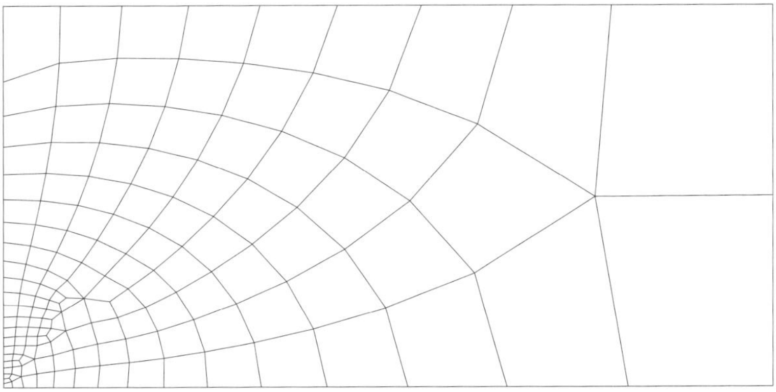

Figure 1.10: Quadrilateral mesh generated using the paving method.

the need for the conversion procedure. Figure 1.10 shows the domain meshed using the paving method. The mesh has 160 elements and 135 nodes. The mesh parameters used are the same as for Figure 1.4.

1.4 Parallel and distributed mesh generation

The size of meshes required for large scale finite element models continues to grow. Their solution may be undertaken using parallel and distributed computing [15, 16, 21,

22, 25, 27]. As the size of finite element problems became larger it was proposed by Khan and Topping [11, 26] that mesh generation should be undertaken in parallel. A number of parallel mesh generators were subsequently developed [14, 17, 23, 26, 29]. It rapidly became apparent that parallel remeshing is related to the mesh partitioning process and these tasks may be integrated as in the sub-domain generation method [12, 18, 24]. For problems that require repeated remeshing it is apparent that dynamic load balancing is required [6]. In a recent paper, Weatherill [13] reports developments in parallel mesh generation motivated by meshes of over 500 million elements.

1.5 Concluding remarks

This preliminary chapter has illustrated the main aspects of generating finite element meshes while providing definitions of the key terms used in mesh generation. In the following chapters each of the main mesh generation methods will be discussed in detail. In the next chapter a discussion of how structured meshes may be generated using the multiblock method, will be presented.

Bibliography

[1] Akin, J.E., *Finite Elements for Analysis and Design*, Academic Press, London, England, 1994.

[2] Babuska, I., Zienkiewicz, O., Gago, J., Oliveira, E.R.d.A., Editors, *Accuracy Estimates and Adaptive Refinements in Finite Element Computations*, John Wiley & Sons, Chichester, England, 1986.

[3] Bathe, K.J., *Finite Element Procedures*, Prentice Hall, Englewood Cliffs, New Jersey, USA, 1996.

[4] Bathe, K.J., *Numerical Methods in Finite Element Analysis*, Prentice-Hall Inc, Englewood Cliffs, New Jersey, USA, 1st Edition, 1978.

[5] Cheung, Y.K., Lo, S.H., Leung, A.Y.T., *Finite Element Implementation*, Blackwell Science, Oxford, England, 1996.

[6] Coupez, T., Marie, S., Ducloux, R., "Parallel 3D simulation of forming processes including parallel re-meshing and reloading", in Desideri, J.A., Le Tallec, P., Onate, E., Periaux, J., Stein, E., Editors, *Numerical Methods in Engineering '96*, pages 738–743. ECCOMAS, John Wiley & Sons, Chichester, England, 1996.

[7] Desai, C.S., Abel, J.F., *Introduction to the Finite Element Method*, Van Nostrand Reinhold Company, New York, USA, 1972.

[8] Frey, P.J., George, P.L., *Mesh generation, application to finite elements*, Hermes Science Publishing, Oxford, UK, 2000.

[9] Hinton, E., Owen, D.J.R., *An Introduction to Finite Element Computations*, Pineridge Press Limited, Swansea, UK, 1979.

[10] Irons, B., Ahmad, S., *Techniques of Finite Elements*, Ellis Horwood, Chichester, England, 1980.

[11] Khan, A.I., Topping, B.H.V., "Parallel Adaptive Mesh Generation", *Computing Systems in Engineering*, 2(3): 75–101, 1991.

[12] Khan, A.I., Topping, B.H.V., "Subdomain Generation for Parallel Finite Element Analysis", *Computing Systems in Engineering*, 4(4-6): 473–488, 1993.

[13] Larwood, B.G., Weatherill, N.P., Hassan, O., Morgan, K., "Domain Decomposition approach for parallel unstructured mesh generation", *Int. J. Numerc. Meth. Engng,*, 58: 177–188, 2003.

[14] Mestres, J.C., Onate, E., Bugeda, G., "PARAMESH: Two Dimensional Unstructured Parallel Mesh Generation Program", Technical Report 61, CIMNE, Barcelona, Spain, March 1995.

[15] Papadrakakis, M., Editor, *Solving Large Scale Problems in Mechanics*, John Wiley & Sons, Chichester, England, 1993.

[16] Papadrakakis, M., Editor, *Parallel Solutions Methods in Computational Mechanics*, John Wiley & Sons, Chichester, England, 1997.

[17] Rypl, D., "Sequential and Parallel Generation of Unstructured 3D Meshes", CTU Report, vol. 2, n. 3, Czech Technical University in Prague, 1998.

[18] Sziveri, J., Seale, C., Topping, B.H.V., "An enhanced parallel sub-domain generation method for mesh partitioning in parallel finite element analysis", *Int. J. Numer. Meth. Eng.*, 47: 1773–1800, 2000.

[19] Thompson, J.F., Soni, B.K., Weatherill, N.P., Editors, *Handbook of Grid Generation*, CRC Press, Boca Raton, USA, 1999.

[20] Thompson, J.F., Warsi, Z.U.A., Mastin, C.W., *Numerical grid generation, Foundation and applications*, Elsevier Science Publishing Co., 1985, also available on Web: http://erc.msstate.edu/education/gridbook/index.html.

[21] Topping, B.H.V., Editor, *Parallel and Distributed Processing for Computational Mechanics: Systems and Tools*, Saxe-Coburg Publications, Edinburgh, UK, 1999.

[22] Topping, B.H.V., Editor, *Computational Mechanics using High Performance Computing*, Saxe-Coburg Publications, Stirling, UK, 2002.

[23] Topping, B.H.V., Cheng, B., "Parallel and Distributed Adaptive Quadrilateral Mesh Generation", *Computers & Structures*, 73 (1-5): 519–536, 1999.

[24] Topping, B.H.V., Khan, A.I., "Domain Decomposition for Parallel Finite Element Computations", in *Parallel Finite Element Computations*, Chapter 10, pages 199–236. Saxe-Coburg Publications, Edinburgh, UK, 1996.

[25] Topping, B.H.V., Khan, A.I., *Parallel Finite Element Computations*, Saxe-Coburg Publications, Edinburgh, UK, 1996.

[26] Topping, B.H.V., Khan, A.I., *Parallel Finite Element Computations*, Chapter 5: Parallel Adaptive Mesh Generation, pages 71–116, Saxe-Coburg Publications, Edinburgh, 1996.

[27] Topping, B.H.V., Lämmer, L., Editors, *High Performance Computing for Computational Mechanics*, Saxe-Coburg Publications, Edinburgh, UK, 2000.

[28] Turner, M.J., Clough, R.W., Martin, H., Topp, L.J., "Stiffness and Deflection Analysis of Complex Structures", *J. Aero. Sci.*, 23: 805–823, 1956.

[29] Wilson, J.K., Topping, B.H.V., "Parallel adaptive tetrahedral mesh generation by

the advancing front technique", *Computers & Structures*, 68(1-3): 57–78, 1998.

[30] Zienkiewcz, O.C., *The Finite Element Method in Structural and Continuum Mechanics*, McGraw Hill, London, 1st Edition, 1967.

[31] Zienkiewicz, O.C., Taylor, R.L., *The Finite Element Method*, Volume 1: The Basis, Chapter 14: Errors, Recovery Processes and Error Estimates, pages 365–400, Butterworth Heinemann, Oxford, England, Fifth Edition, 2000.

[32] Zienkiewicz, O.C., Taylor, R.L., *The Finite Element Method*, Volume 1: The Basis, Chapter 15: Adaptive Finite Element Refinement, pages 401–428, Butterworth Heinemann, Oxford, England, Fifth Edition, 2000.

Chapter 2

Block structured model and mesh generation

2.1 Introduction

A *structured mesh* consists of two sets of lines in two dimensions or three sets of lines in three dimensions. In two dimensions the lines in the same set do not intersect each other and they intersect the lines in the other sets only once. These meshes are also known as *regular grids* since the connectivity of the elements is regular. In the generation of these meshes the basic concept is that a structured grid is generated in a simple "logical" domain (such as a rectangle or a triangle) and this grid is mapped to the real domain, also ensuring boundary compatibility.

The basic methods used to generate structured meshes as classified by Frey and George [4] are:

- algebraic interpolation methods,
- solving partial differential equations, and
- multiblock.

This classification is not exclusive and in most cases some combinations are used. For a detailed and rigorous presentation of the methods see also Thompson *et al* [15] which is also available on the Internet.

In the following sections a multiblock method with the algebraic interpolation is discussed.

2.2 Overview of the multiblock method

The multiblock method is based on an approach where the domain to be meshed is partitioned into elementary geometric entities. After the decomposition, the elemen-

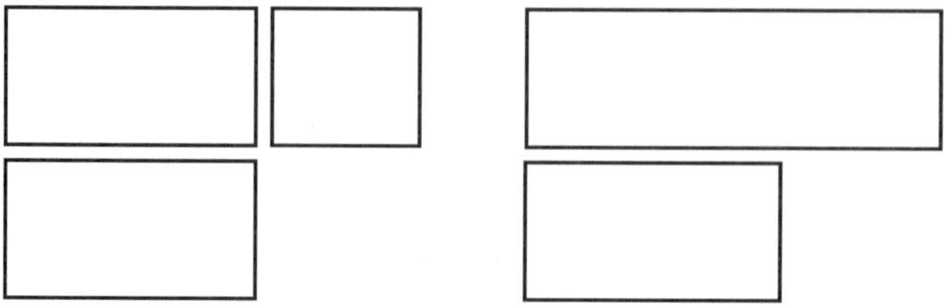

(a) (b)

Figure 2.1: (a) Conforming and (b) non-conforming division of the domain into blocks.

tary blocks are remeshed using a local meshing process (such as an algebraic method) and finally they are merged back into one domain.

One of the advantages of the method is that it can easily provide boundary fitting grids using curvilinear coordinates and that the spacing of the grid points (even internal points) is directly controllable. Furthermore the computational code required for structured grids can be simpler than that for unstructured grids since the program can simply go through all elements in loops and at the same time the adjacency is implicitly provided. As the original domain is divided into independent blocks it also provides a direct way for the parallelisation of not only the mesh generation but the computational code. The program described in this book was developed initially for cable membrane structures [16].

The method can be divided into the following basic steps:

1. **Partitioning of the domain**

   The automatic division or decomposition of structures into elementary geometric blocks is one of the most important aspects of this method. When it is completed manually it can be a difficult and labour intensive task. Several researchers have attempted to automate this process or devise rules for the decomposition (see references [1, 12]), while in reference [10] a fully automatic method for decomposition is reported.

   The main objective to achieve at this step is to cut the domain into a small number of convex blocks. (The convexity of quadrilateral blocks is especially important to ensure that all nodes and elements will be generated inside and that the quality of the elements will be acceptable.) The partitioning of the domain can be conforming or non-conforming as shown in Figure 2.1. In the case of non-conforming partitioning special attention should be paid to ensuring compatibility on the shared edges.

   The geometric "blocks" in 2D may be surfaces formed from lines, and quadrilateral or triangular shapes. This information can be supplied in various ways, but in the implementation described in this book a geometry data file is used. This file should be created manually and no automatic decomposition is provided. The syntax of the file is described in Appendix B.

2. **Division of the blocks**

   The geometric blocks are divided into smaller elements. The method used is described in detail in Section 2.3.

3. **Merging of the blocks**

   Finally the geometric blocks are merged together. The compatibility of the interfaces of the blocks is either ensured explicitly after the mesh generation or implicitly by the method itself. Some discussion on the merging of the blocks is presented in Section 2.4.

2.3 Meshing of the surface geometric blocks

The blocks of the surface geometric entities can be meshed by an algebraic grid generator [5, 13] or by solving partial differential equations [3, 14, 15].

The advantage of the algebraic grid generator is its simplicity and speed compared to solving partial differential equations. The algebraic grid generator generates a mesh in a logical space domain and then maps all points back to the real domain. Using the appropriate mapping function boundary conformity can be achieved. The main decision that has to be made with this method is to find an appropriate mapping function between the logical and the real domain. One of the most well-known mapping functions is obtained via transfinite interpolation (as described in references [6–8]).

The algebraic method with the transfinite mapping can be described by following George [5] for a block domain Ω with a reasonable quadrilateral shape (see Figure 2.2). In this case reasonable quadrilateral shape means that the internal angles of the block should each be approximately 90 degrees. It is assumed that on the opposite sides of the quadrilateral shapes the number of points should be equal, however their distribution need not be uniform. If $\overline{\Omega}$ is a unit square, the number of points on the four sides of the unit square should be equal to the number of the points on the sides of the real domain and their distribution should correspond to the real point distribution. This can be seen as phase 1 in Figure 2.2. To obtain the *reference mesh* connectivity on the $\overline{\Omega}$ domain the corresponding points on the opposite sides are connected. The intersection of the connecting lines determines the position of the internal points. These points can be seen in phase 2 in Figure 2.2. Finally, a function F maps all the points of the $\overline{\Omega}$ domain to the domain of Ω which is phase 3 in Figure 2.2. Using the same connectivity of $\overline{\Omega}$ in domain Ω the elements of the domain Ω are defined. The robustness of the method and its capability to handle all types of geometric problems depends on the properties of the function F.

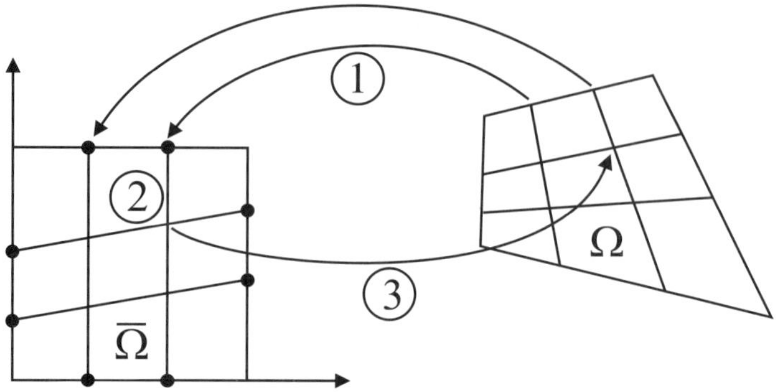

Figure 2.2: Algebraic method with mapping.

However in this implementation phases 2 and 3 are merged by using an "interpolation function". In this case the generation of points is as follows. If the local coordinate system is (ξ, η), and the range of ξ and η is between 0 and 1 then the number of subdivisions in the ξ direction is defined by n_ξ and in the η direction by n_η. The weighting factors of the subdivisions are defined by w_ξ and w_η and they express the relative length of the subdivisions to each other. The coordinates of the points can be calculated as follows:

1. Both coordinate values (ξ and η) are initialised as 0.
2. Calculate the coordinates ($x(\xi,\eta), y(\xi,\eta)$ and $z(\xi,\eta)$) of all nodes on the surface using an "interpolation function".
3. Increment the local coordinates by

$$\xi_i = \xi_{i-1} + \frac{(w_\xi)_i}{w_\xi^{total}} \tag{2.1}$$

$$\eta_i = \eta_{i-1} + \frac{(w_\eta)_i}{w_\eta^{total}} \tag{2.2}$$

where

$$w_\xi^{total} = \sum_{i=1}^{n_\xi} (w_\xi)_i, \quad \text{and} \quad w_\eta^{total} = \sum_{i=1}^{n_\eta} (w_\eta)_i \tag{2.3}$$

4. If the element generation is not finished repeat from step 2.

The method provides quadrilateral connectivity initially. The triangular elements are obtained from the quadrilateral elements by connecting two diagonal points of each of the quadrilateral elements.

The simplest "interpolation function" allows linear approximation of the sides of the blocks. For the linear approximation only the two endpoints of the sides should be specified. For quadratic approximation middle side nodes are required to define the curved boundary. This approximation was used in the work of Hinton and Owen [9]. A more practical approximation is a parametric cubic curve or surface approximation, since this type of geometric description is used in state-of-the-art CAD systems [2,11].

In the program described in this book the available geometric entities are straight lines, quadrilateral cubic surfaces with curved sides and triangular linear surfaces with straight sides.

The node distribution on the sides may be uniform or nonuniform. In the case of the quadrilateral block the number of nodes generated on the opposite sides must be equal.

2.3.1 Meshing of triangular surfaces with linear interpolation

The meshing of triangular surfaces with straight edges is an easy task. The only difficulty arises when the distribution of the points on the edges is not uniform. One solution is to use the transfinite mapping method for triangles as suggested by George [5]. In the implementation presented here an alternative approach is used to generate the internal points.

When the distribution of the points on all sides is uniform all internal points are placed at the intersection of three lines which are parallel with the three edges. This situation is shown in Figure 2.3(a), where the intersection point is circled. When the distribution of the points on the sides is non-uniform then the three lines are not parallel with the corresponding sides and they intersect each other pairwise. These intersection points determine a triangular shape which is shown in Figure 2.3(b) as a shaded area. The new internal point will be generated at the centre of this triangular shape. An example of mesh generation in a triangular shape with non-uniform point distribution on the edges is shown in Figure 2.4.

2.3.2 Meshing of quadrilateral surfaces with cubic interpolation

By considering four parametric cubic curves in space connected at their end nodes a surface cubic patch can be defined as shown in Figure 2.5. The four cubic curves require 4 by 4, or 16 parameters. The algebraic format of a patch surface is

$$P(\xi,\eta) = \left\{ \begin{array}{cccc} 1 & \xi & \xi^2 & \xi^3 \end{array} \right\} \begin{bmatrix} S_{00} & S_{01} & S_{02} & S_{03} \\ S_{10} & S_{11} & S_{12} & S_{13} \\ S_{20} & S_{21} & S_{22} & S_{23} \\ S_{30} & S_{31} & S_{32} & S_{33} \end{bmatrix} \left\{ \begin{array}{c} 1 \\ \eta \\ \eta^2 \\ \eta^3 \end{array} \right\} \tag{2.4}$$

or

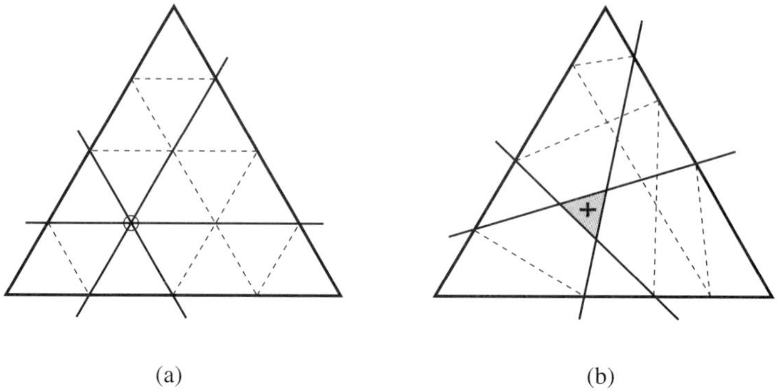

Figure 2.3: Internal point generation in the case of (a) uniform (◯) and (b) non-uniform (+) point distribution on the sides.

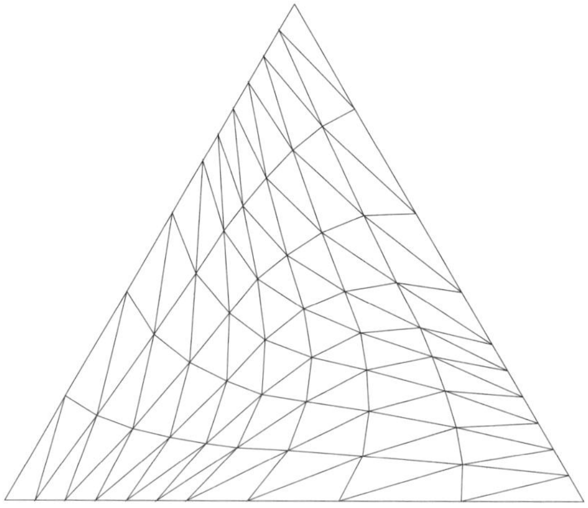

Figure 2.4: Example of the mesh generation in a triangular entity with non-uniform point distribution on the sides.

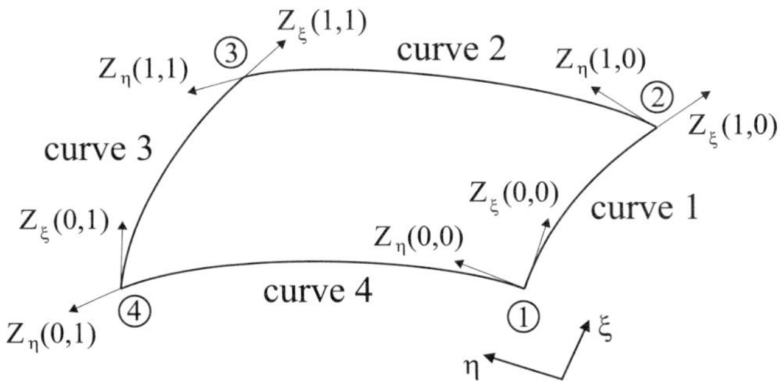

Figure 2.5: Cubic patch definition using four curves and derivatives.

$$P(\xi,\eta) = \sum_{i=0}^{3}\sum_{j=0}^{3} S_{ij}\,\xi^i\,\eta^j \tag{2.5}$$

where P is a general coordinate component. For a three dimensional patch there are three components representing x, y and z. The parametric variables ξ and η are defined in the region of $0 \le \xi \le 1$ and $0 \le \eta \le 1$. One way to determine the unknown S_{ij} parameters is to use the derivative forms of the parametric cubic patch. In this case the expanded format of equation 2.4 is:

$$\begin{aligned} P(\xi,\eta) = {} & S_{00} && + S_{01}\eta && + S_{02}\eta^2 && + S_{03}\eta^3 \\ + {} & S_{10}\xi && + S_{11}\xi\eta && + S_{12}\xi\eta^2 && + S_{13}\xi\eta^3 \\ + {} & S_{20}\xi^2 && + S_{21}\xi^2\eta && + S_{22}\xi^2\eta^2 && + S_{23}\xi^2\eta^3 \\ + {} & S_{30}\xi^3 && + S_{31}\xi^3\eta && + S_{32}\xi^3\eta^2 && + S_{33}\xi^3\eta^3 \end{aligned} \tag{2.6}$$

The derivative of $P(\xi,\eta)$ with respect to ξ is given by $P_\xi(\xi,\eta)$:

$$\begin{aligned} P_\xi(\xi,\eta) = {} & 0 && + 0 && + 0 && + 0 \\ + {} & S_{10} && + S_{11}\eta && + S_{12}\eta^2 && + S_{13}\eta^3 \\ + {} & 2S_{20}\xi && + 2S_{21}\xi\eta && + 2S_{22}\xi\eta^2 && + 2S_{23}\xi\eta^3 \\ + {} & 3S_{30}\xi^2 && + 3S_{31}\xi^2\eta && + 3S_{32}\xi^2\eta^2 && + 3S_{33}\xi^2\eta^3 \end{aligned} \tag{2.7}$$

The derivative of $P(\xi,\eta)$ with respect to η is given by $P_\eta(\xi,\eta)$:

$$\begin{aligned} P_\eta(\xi,\eta) = {} & 0 && + S_{01} && + 2S_{02}\eta && + 3S_{03}\eta^2 \\ + {} & 0 && + S_{11}\xi && + 2S_{12}\xi\eta && + 3S_{13}\xi\eta^2 \\ + {} & 0 && + S_{21}\xi^2 && + 2S_{22}\xi^2\eta && + 3S_{23}\xi^2\eta^2 \\ + {} & 0 && + S_{31}\xi^3 && + 2S_{32}\xi^3\eta && + 3S_{33}\xi^3\eta^2 \end{aligned} \tag{2.8}$$

The derivative of $P(\xi,\eta)$ with respect to ξ and η is given by $P_{\xi\eta}(\xi,\eta)$:

$$\begin{aligned} P_{\xi\eta}(\xi,\eta) = \; & 0 && +0 && +0 && +0 \\ & +0 && +S_{11} && +2S_{12}\eta && +3S_{13}\eta^2 \\ & +0 && +2S_{21}\xi && +4S_{22}\xi\eta && +6S_{23}\xi\eta^2 \\ & +0 && +3S_{31}\xi^2 && +6S_{32}\xi^2\eta && +9S_{33}\xi^2\eta^2 \end{aligned} \tag{2.9}$$

If four pairs of values, $(0,0)$, $(1,0)$, $(0,1)$ and $(1,1)$, for ξ and η are substituted into Equations 2.6-2.9, then the resulting equation in matrix form is

$$\begin{Bmatrix} P(0,0) \\ P(1,0) \\ P(1,1) \\ P(0,1) \\ P_\xi(0,0) \\ P_\xi(1,0) \\ P_\xi(1,1) \\ P_\xi(0,1) \\ P_\eta(0,0) \\ P_\eta(0,1) \\ P_\eta(1,1) \\ P_\eta(1,0) \\ P_{\xi\eta}(0,0) \\ P_{\xi\eta}(0,1) \\ P_{\xi\eta}(1,1) \\ P_{\xi\eta}(1,0) \end{Bmatrix} = [\mathbf{M}] \begin{Bmatrix} S_{00} \\ S_{01} \\ S_{02} \\ S_{03} \\ S_{10} \\ S_{11} \\ S_{12} \\ S_{13} \\ S_{20} \\ S_{21} \\ S_{22} \\ S_{23} \\ S_{30} \\ S_{31} \\ S_{32} \\ S_{33} \end{Bmatrix} \tag{2.10}$$

or in short form

$$\{\mathbf{P}_{der}\} = [\mathbf{M}]\{\mathbf{S}\} \tag{2.11}$$

where $\mathbf{M}$ is a 16 by 16 coefficient matrix. If the $\mathbf{P}_{der}$ vector is known then Equation 2.11 can be solved for $\mathbf{S}$

$$\{\mathbf{S}\} = \left[\mathbf{M}^{-1}\right]\{\mathbf{P}_{der}\} \tag{2.12}$$

which, after rearrangement of the resulting vector $\{\mathbf{S}\}$ into a matrix form $[\mathbf{S}]$, can be substituted into Equation 2.4 :

$$P(\xi,\eta) = \{\; 1 \quad \xi \quad \xi^2 \quad \xi^3 \;\}\; [\mathbf{S}] \begin{Bmatrix} 1 \\ \eta \\ \eta^2 \\ \eta^3 \end{Bmatrix} \tag{2.13}$$

A cubic patch can also be defined uniquely by 16 points as shown in Figure 2.6. In this case, for example Equation 2.13 can be used where the values of $P(\xi, \eta)$ are known (16 of them, one for each point)

$$\left\{ \begin{array}{c} P(\xi_0, \eta_0) \\ P(\xi_1, \eta_1) \\ P(\xi_2, \eta_2) \\ \cdots \\ P(\xi_{15}, \eta_{15}) \end{array} \right\} = [\mathbf{N}] \left\{ \begin{array}{c} P(0,0) \\ P(1,0) \\ P(1,1) \\ \cdots \\ P_{\xi\eta}(1,0) \end{array} \right\} \tag{2.14}$$

or in short form

$$\{\mathbf{P}_{pos}\} = [\mathbf{N}] \ \{\mathbf{P}_{der}\} \tag{2.15}$$

and this newly formed equation is solved for P_{der}. Next P_{der} can be substituted back to Equation 2.12 and after rearrangement Equation 2.13 can be solved again. An alternative way is, for example, using Equation 2.4 with the given 16 points:

$$\left\{ \begin{array}{c} P(\xi_0, \eta_0) \\ P(\xi_1, \eta_1) \\ P(\xi_2, \eta_2) \\ \cdots \\ P(\xi_{15}, \eta_{15}) \end{array} \right\} = [\mathbf{K}] \left\{ \begin{array}{c} S_{00} \\ S_{01} \\ S_{02} \\ \cdots \\ S_{33} \end{array} \right\} \tag{2.16}$$

Then the same method can be used as in Equation 2.12 and 2.13.

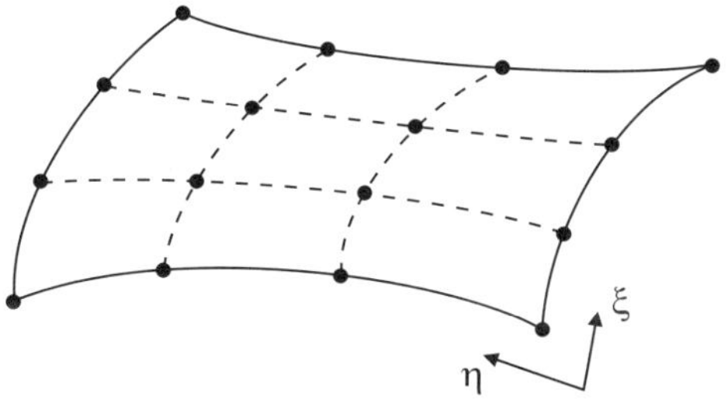

Figure 2.6: Cubic patch definition with 16 points.

2.4 Common boundaries

Since the geometric entities are meshed individually the generated points and elements are numbered separately. Points on the common boundaries between entities have the same coordinate values, but they will generally have different numbers. These duplicate points should be eliminated and finally only one point should be on the boundary with a unique point or node number. The removal of the duplicate points ensures the creation of a conforming finite element mesh.

One way to remove the duplicate points is to search all points systematically and compare their coordinate values to identify the equal points [9]. For example, if point 14 and point 211 have the same coordinate values then point 211 is deleted. In the element connectivity data all reference to point 211 is replaced with a reference to point 14. Furthermore to ensure the continuous numbering of the points, all points with higher number than 211 should be renumbered (decreased by one) and all references to these points should be updated at the same time. Unfortunately the time of the systematic search grows "exponentially" as the number of generated points increases. It is also important to note that for this method, it is crucial to ensure that the subdivisions of the boundary edges of the adjacent entities are compatible, otherwise no duplicate points will be found and the blocks would form separate meshes.

To avoid these problems George [5] suggested a method which requires the creation of the list of the edges of the initial geometric entities. Using this list of edges each edge is refined only once and the generated points are numbered uniquely. The creation order of the points on a particular edge is determined by the directional sense of the edge definition, for example the endpoint with a smaller identification number will be used as a starting point in the generation process. Once the points on the edges have been generated the internal points of the blocks can be generated and unique numbers can be assigned to each newly created point sequentially.

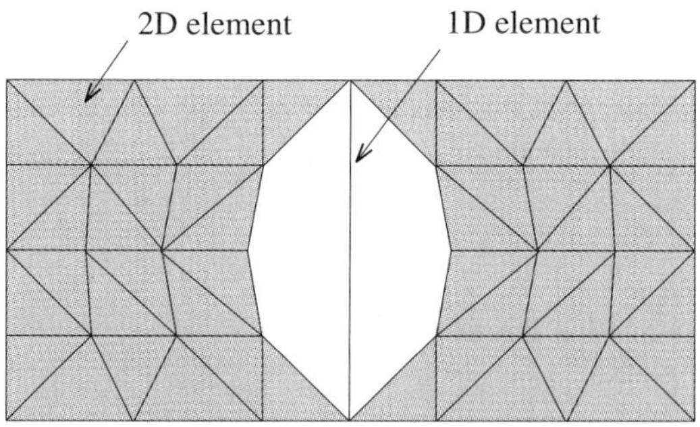

Figure 2.7: Generation of 2D and 1D elements when they are attached to each other at two points and not along the side.

The approach of searching for duplicate nodes by comparison of nodal coordinates has been adopted. The reason is that it should be possible to generate a mesh as shown in Figure 2.7 where the membrane and cable elements are attached to each other at only two nodes and not along the side. When using the method to remove duplicate nodes suggested by George [5] extra care and specification would be required to allow for situations of the type shown in Figure 2.7.

2.5 Concluding remarks

In this chapter a method for generating structured meshes based on algebraic interpolation has been presented. In the next chapter a framework for using methods for generating unstructured meshes will be discussed. Procedures for adaptively controlling the mesh size using mesh parameters will also be described.

Bibliography

[1] Allwright, S.E., "Techniques in multiblock domain decomposition and surface grid generation", in Sengupta, S., Thompson, J.F., Eiseman, P.R., Hauser, J., Editors, *Numerical Grid Generation in Computational Fluid Mechanics '88*, pages 559–568. Pineridge Press, Swansea, UK, 1988.

[2] Association, N.C.G., "Initial Graphics Exchange Specification (IGES) Version 5.3", Technical report, IGES/PDES Organization, 2722 Merrilee Drive, Suite 200, Fairfax, VA., USA, 1995.

[3] Chan, W.M., "Hyperbolic methods for surface and field grid generation", in Thompson, J.F., Soni, B.K., Weatherill, N.P., Editors, *Handbook of Grid Generation*, Chapter 5, pages 1–26. CRC Press, Boca Raton, USA, 1999.

[4] Frey, P.J., George, P.L., *Mesh generation, application to finite elements*, Chapter 4, "Algebraic, P.D.E. and multiblock methods", pages 135–164, Hermes Science Publishing, Oxford, England, 2000.

[5] George, P.L., *Automatic Mesh Generation, Application to Finite Element Methods*, John Wiley & Sons, Chichester, England, 1991.

[6] Gordon, W.J., Hall, C.A., "Construction of curvilinear co-ordinate systems and applications to mesh generation", *International Journal for Numerical Methods in Engineering*, 7: 461–477, 1973.

[7] Haber, R., Abel, J.F., "Discrete transfinite mappings and meshing of three-dimensional surfaces using interactive computer graphics", *International Journal for Numerical Methods in Engineering*, 18: 41–66, 1982.

[8] Haber, R., Shephard, M.S., Greenberg, D.P., "A general two-dimensional graphical finite element preprocessor utilizing discrete transfinite mappings", *International Journal for Numerical Methods in Engineering*, 17: 1015–1044, 1981.

[9] Hinton, E., Owen, D.R.J., *Finite element programming*, Academic Press, London, England, 1977.

[10] Park, S., Lee, K., "A new approach to automated multiblock decomposition for grid generation: A Hypercube++ approach", in Thompson, J.F., Soni, B.K., Weatherill, N.P., Editors, *Handbook of Grid Generation*, Chapter 10, pages 1–22. CRC Press, Boca Raton, USA, 1999.

[11] PDA Engineering, PATRAN Division, 2975 Redhill Avenue, Costa Mesa, California, USA, *PATRAN Plus User Manual*, October 1990.

[12] Shaw, J.A., Weatherill, N.P., "Automatic topology generation for multiblock grids", *Applied Mathematics and Computation*, 53: 355–388, 1992.

[13] Smith, R.E., "Transfinite Interpolation (TFI) Generation Systems", in Thompson, J.F., Soni, B.K., Weatherill, N.P., Editors, *Handbook of Grid Generation*, Chapter 3, pages 1–15. CRC Press, 1999.

[14] Spekreijse, S.P., "Elliptic generation systems", in Thompson, J.F., Soni, B.K., Weatherill, N.P., Editors, *Handbook of Grid Generation*, Chapter 4, pages 1–49. CRC Press, Boca, Raton, USA, 1999.

[15] Thompson, J.F., Warsi, Z.U.A., Mastin, C.W., *Numerical grid generation, Foundation and applications*, Elsevier Science Publishing Co., 1985, also available on Web: http://erc.msstate.edu/education/gridbook/index.html.

[16] Topping, B.H.V., Iványi, P., *Computer Aided Design of Cable-Membrane Structures*, Saxe-Coburg Publications, Stirling, UK, 2004.

Chapter 3

Unstructured mesh generation

3.1 Introduction

Without doubt the three most important two-dimensional unstructured mesh generation methods are:

- The advancing front method [5–7] as discussed in Chapter 4.
- The paving method [1, 2] as discussed in Chapter 5, and
- Delaunay triangulation based mesh generation [4, 8] discussed in Chapters 6 and 7.

The advancing front and the Delaunay mesh generation methods are particularly suited for generating meshes consisting of triangles (in 2D) or tetrahedra (in 3D). The paving method provides an excellent tool for constructing meshes of quadrilateral elements.

3.2 Mesh generator layout

The three methods of interest can be implemented in a shared framework of general tools and input/output (I/O) functions. The flowchart for the integrated multi-method unstructured mesh generator, developed by the authors, is shown in Figure 3.1.

◦ 1. **load & parse job description**

The program loads a file which contains all instructions that indicate what kind of meshing operation has to be carried out and on which files. This description of the job is analysed to check for inconsistencies, unimplemented options, or missing information. Appendix C describes the syntax and internal structure of this job description as used by the authors.

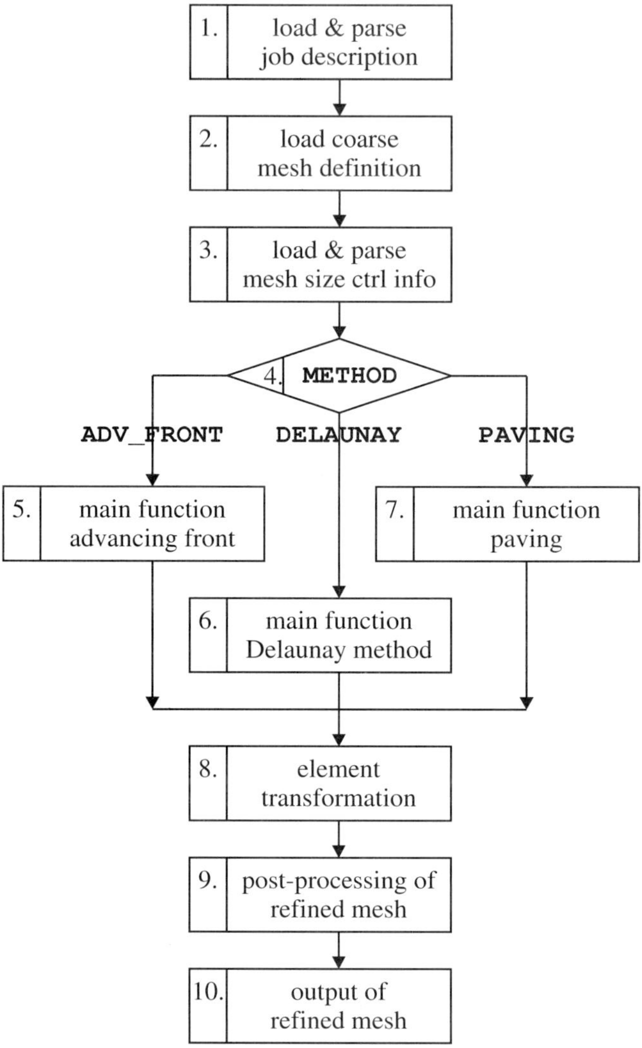

Figure 3.1: Unstructured mesh generator flowchart.

∘ 2. **load coarse mesh definition**

From a file of which the name is mentioned in the job description a coarse mesh is loaded into the memory in the form of an E_Library mesh definition data structure. For details of the E_Library see Appendix D. This loading process includes the validity checking of the mesh definition. Routine checks are performed such as searching for the presence of duplicate nodes or invalid definitions of elements.

○ 3. **load & parse mesh size control information**

As indicated in the job description, all files related to the control of the mesh size will be loaded and parsed here. The different concepts of mesh size control are described in section 3.3 and this step will be further illustrated by the flowchart shown in Figure 3.8.

○ 4. **method selection**

The job description indicates which mesh generation method is chosen by the user for executing the mesh generation job.

○ 5. **the advancing front method**

If the advancing front method explained in Chapter 4 is chosen, the program layout shown in Figure 4.1 follows.

○ 6. **the Delaunay triangulation method**

If the Delaunay triangulation method explained in Chapter 6 is chosen, the program layout shown in Figure 6.5 follows.

○ 7. **the paving method**

If the paving method, explained in Chapter 5, is chosen the program layout shown in Figure 5.1 follows.

○ 8. **element transformation**

In order to give the user maximum flexibility, procedures are introduced to convert a mesh consisting of triangles into a mesh made of quadrilaterals and *vice versa*. This subject is discussed in greater detail in Chapter 8.

○ 9. **post-processing of the fine mesh**

After the mesh refinement procedure the resulting fine mesh will be smoothed and its quality enhanced according to the method chosen. This step is the subject of Chapter 9.

○ 10. **output of the fine mesh**

The output of the newly generated mesh in the E_Library mesh definition format concludes the mesh generation job. The name of the output file is recorded in the mesh generation job description.

3.3 Mesh size control

In this section the use of sources [3] to control the mesh sizes are explained. These sources are used to specify the mesh parameters that control the mesh sizes in refined meshes. The domain to be meshed is usually provided as an *initial or coarse mesh.*

The coarse mesh specifies the coordinates of the meshpoints and the topology of the initial elements. These elements can be triangles or quadrilateral elements. The coarse mesh provides the information about the final shape of the domain. It may also specify how the boundary and the internal areas of the mesh should be refined.

To control the size of the generated elements in various parts of the coarse mesh three methods are presented. Each method has its own typical application field.

3.3.1 Mesh parameters

The key concept of the size control of the elements is to specify the maximum element sizes which the elements in the refined mesh should not exceed. However a finer level of control can be reached when maximum edge length sizes are introduced. The maximum edge length is expressed as a mesh parameter. The mesh parameter can be assigned to elements or meshpoints. The implementation described in this book relies mainly on nodal mesh parameters, therefore in the case of element mesh parameters all mesh parameters of elements which are connected to a node are averaged and the average mesh parameter is assigned to the particular node.

The nodal mesh parameter generally expresses that any edge connected to the meshpoint is no longer than the specified mesh parameter. In this way, each end of an edge in the coarse mesh will have an assigned mesh parameter and upon refinement the edge will then be split up into smaller segments so that each segment is no longer than the value of the mesh parameter assigned to any of its end nodes.

The procedure to determine the coordinates of the intermediate nodes and the values assigned to these nodes are discussed in Appendix A. Further considerations for the generation of the boundary meshpoints are discussed in the chapter corresponding to the particular method.

Mesh parameters do not have to be assigned directly by the user. A preliminary analysis and error estimation could be carried out on a coarse discretisation of the domain for the purpose of determining the required mesh sizes. These techniques of converting error values to mesh size requirements are essential for adaptive mesh generation.

3.3.2 Background mesh

The maximum element or edge size at a certain location in the domain can also be derived by looking up the corresponding value of that location in a second mesh. This second mesh is referred to as a *background mesh*. It extends over the entire mesh domain and can be imagined as lying on top of the mesh domain as a transparent domain or field. A certain location in the coarse mesh belongs at the same time to an element of the background mesh. This location then inherits the value of the background element's mesh parameter.

The concept of nodal mesh parameters can also be applied to the background mesh in which case the background and coarse mesh will be the same mesh. The maximum edge size at a certain location will then be derived by interpolating between the nodal background mesh parameters of the corresponding background element corner nodes. Interpolation schemes such as area weighting can be applied. Figure 3.2 demonstrates this interpolation.

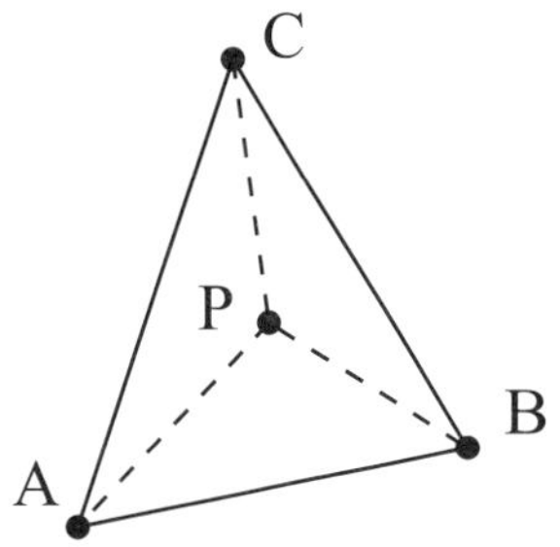

$\delta_P = \frac{\delta_A \triangle_{BCP} + \delta_B \triangle_{CAP} + \delta_C \triangle_{ABP}}{\triangle_{ABC}}$ where δ_i stands for the mesh size attached to node i and $\triangle_{ABC}$ is the area of triangle ABC.

Figure 3.2: Area weighted interpolation of nodal mesh parameters.

3.3.3 Local control sources

For some types of finite element analysis such as CFD, precise local mesh control is important because of the special nature of the movement of high speed fluid. For this purpose local control sources were developed [3] as an extension of the background mesh control method. First the required mesh parameter is obtained from the background mesh. This parameter may be modified in the vicinity of locally defined areas. These areas are called local control sources. Two types of local control sources will be defined here:

- point sources, and
- line sources.

3.3.3.1 Point sources

A point source for local mesh control is defined by four parameters as shown in Figure 3.3:

- the coordinates of the centre of the point source $p(x_p, y_p)$,
- an inner radius r_i,
- an outer radius r_o,

- and a core mesh parameter δ_s.

Assuming that the mesh parameter derived from the background for location $l(x, y)$ is δ_{bg} and $d(l, p)$ is the distance from the centre of the point source to the location, the modified mesh δ_m parameter through the influence of the source may be computed using linear interpolation as:

$$\delta_m = \begin{cases} \delta_s & \text{if } d(l,p) < r_i \\ \delta_{bg} & \text{if } d(l,p) > r_o \\ \delta_s + \frac{d(l,p) - r_i}{r_o - r_i}(\delta_{bg} - \delta_s) & \text{else} \end{cases} \tag{3.1}$$

As seen in Figure 3.4 the point source provides a local zone where a different mesh size with a certain radius can be applied. A transition towards the background value is made in a ring around the zone.

3.3.3.2 Line sources

Line sources are a similiar concept to point sources. The local zone of different mesh size is extended from a circle to a rounded zone around a poly-line. A poly-line is a series of connected lines where the end of one line is the start of another line. A line source is defined by a series of point sources that are linked.

Figure 3.5 shows the areas in which different interpolation schemes apply. If the location for which a modified parameter is required falls outside of the line source influence area (marked by a strong outer boundary in the figure) the background mesh parameter is returned unmodified. For a location in zone A the core mesh parameter of the respective point source is returned. In zone B the interpolation procedure of Equation 3.1 is applied for the respective point. Zones C and D provide a smooth continuous transition between the schemes applied for the two point sources.

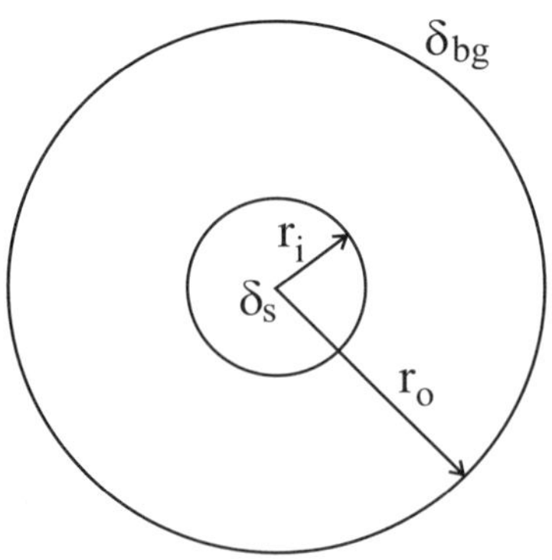

Figure 3.3: Point source interpolation.

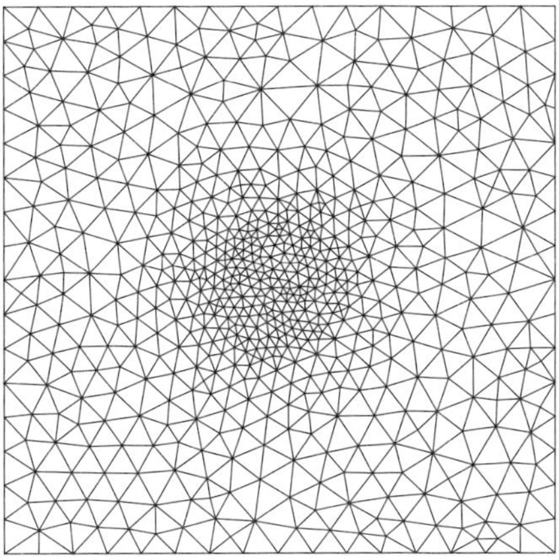
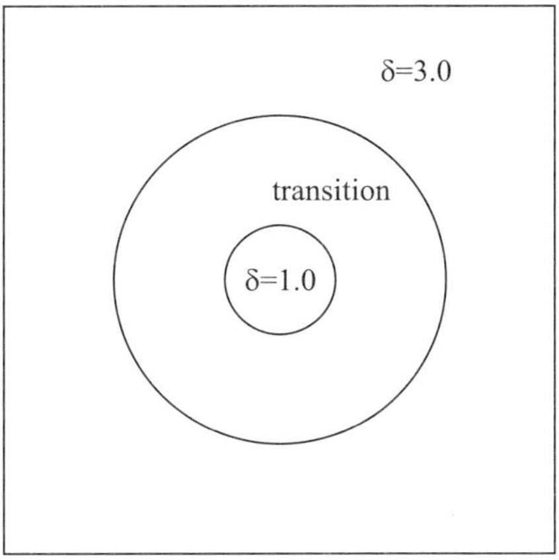

Figure 3.4: An example point source for local mesh size control.

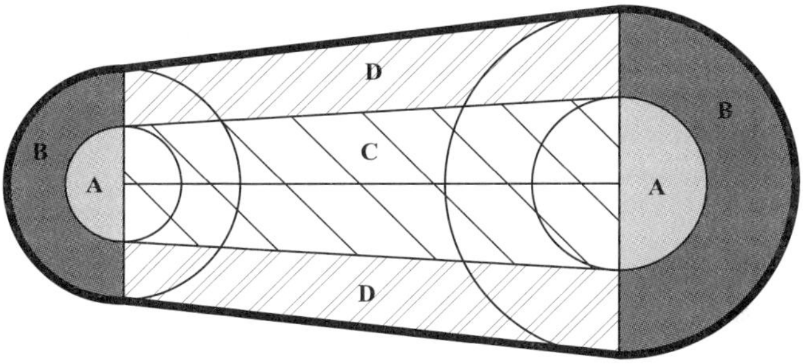

Figure 3.5: A line source for local mesh size control.

Figure 3.6 shows the layout resulting in the following interpolation equations. For locations in zone C such as point p_2 the modified mesh parameter is given by linear interpolation as:

$$\delta_{m2} = \delta_{s1} + (\delta_{s2} - \delta_{s1}) * \frac{d_2}{d_s} \tag{3.2}$$

Therefore the modified mesh parameter for locations in zone C depends uniquely on the projection of the location on the line that connects the centres of the sources.

For locations in zone D more computation is required. First the width of the line source's inner core r_a and outer boundary r_b at the location must be determined using

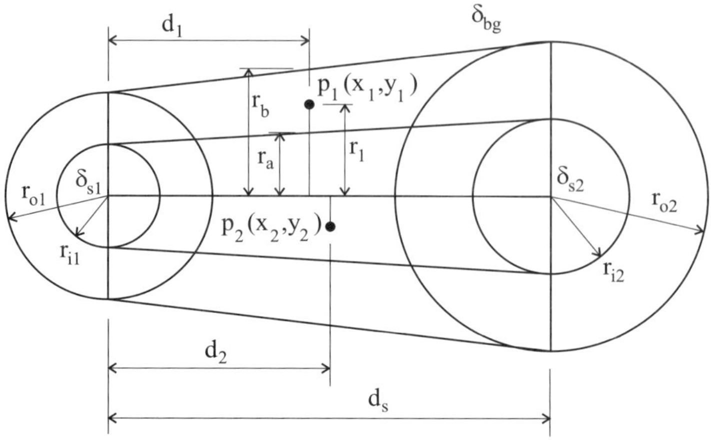

Figure 3.6: Line source interpolation.

linear interpolation as follows:

$$r_a = r_{i1} + (r_{i2} - r_{i1}) * \frac{d_1}{d_s} \tag{3.3}$$

$$r_b = r_{o1} + (r_{o2} - r_{o1}) * \frac{d_1}{d_s} \tag{3.4}$$

Next the value δ^*_{m1} of the modified mesh parameter that would have been assigned if the location lay on the same perpendicular distance from but inside zone C is calculated.

$$\delta^*_{m1} = \delta_{s1} + (\delta_{s2} - \delta_{s1}) * \frac{d_1}{d_s} \tag{3.5}$$

The final interpolation yields the modified mesh parameter as

$$\delta_{m1} = \delta^*_{m1} + (\delta_{bg} - \delta^*_{m1}) * \frac{r_1 - r_a}{r_b - r_a} \tag{3.6}$$

Finally an example for the line source interpolation is shown in Figure 3.7.

3.3.3.3 Data structures for local mesh control

In the implementation discussed in this book two data structures were created to cater for the information relating to the local mesh control: `PARAMCNTL` and `PARAMCNTLSETS`.

The data struture `PARAMCNTL` defines the information for each local mesh control data set. It contains the number of control points for this data set. It also specifies the inner radius, outer radius, point coordinates and mesh size parameter for each control

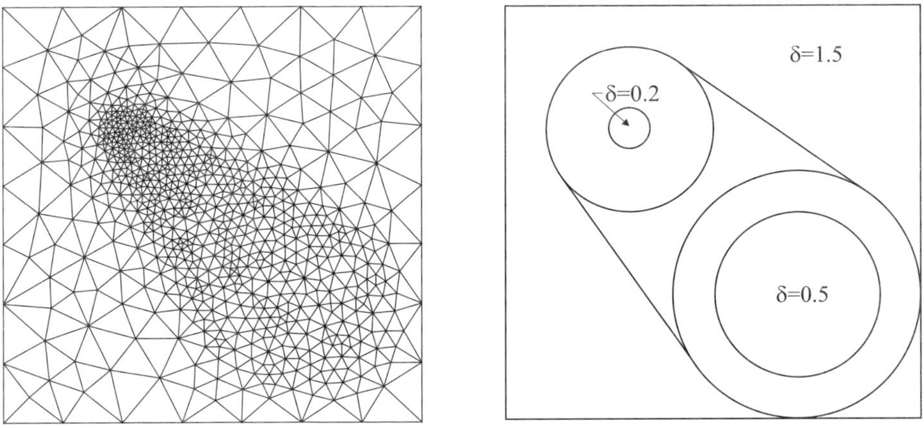

Figure 3.7: An example line source for local mesh size control.

point. Obviously the point-wise local control is treated as a special case of line-wise local control. It is unnecessary to create an extra data structure for point-wise control.

```
typedef struct
{
   int NCntlPoints;
   double *InnerRadius;
   double *OuterRadius;
   double **Coords;
   double *Param;
}
PARAMCNTL;
```

The data structure `PARAMCNTLSETS` encapsulates the general information about a number of local mesh control sets and their corresponding memory addresses.

```
typedef struct
{
   int NCntlSets;
   PARAMCNTL *ParamCntls;
}
PARAMCNTLSETS;
```

3.3.3.4 File format

The file format of the local control sources is discussed in Appendix C.

3.3.4 Implementation of refinement functions

The procedures for creating new refinement nodes on line segments between existing points are implemented using the functions detailed in Appendix A.

3.3.5 Loading of the mesh size control information

Figure 3.8 shows a detailed flowchart of the actions to be taken related to the loading and parsing of files that control the mesh size over the domain when the refinement procedure starts.

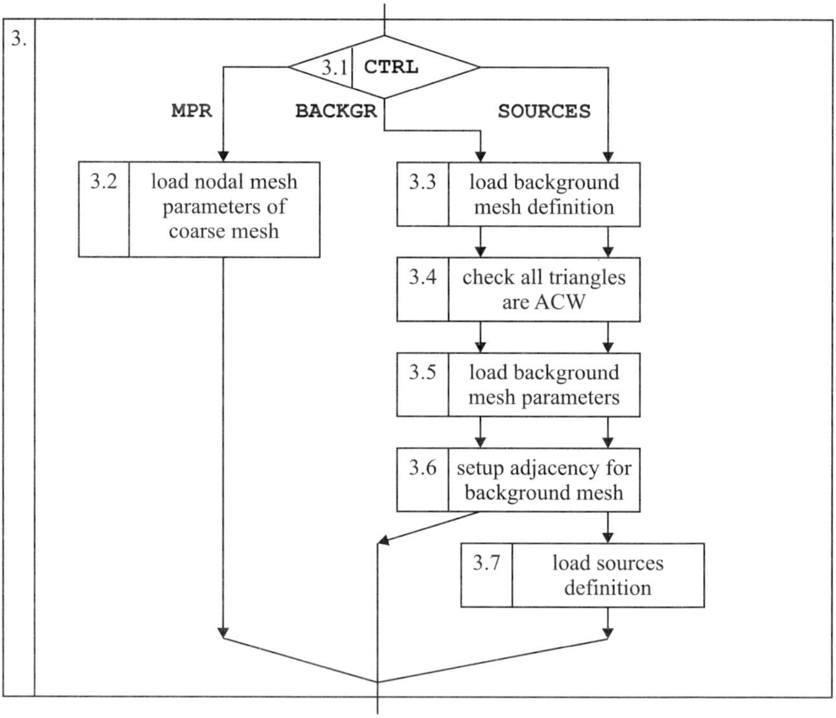

Figure 3.8: Flowchart for loading and parsing of the mesh size control data.

○ 3.1 **ctrl selection**

Different actions are taken in this procedure depending on which method is selected for the mesh size control. The possible options are `MPR` (mesh parameter control see section 3.3.1), `BACKGR` (background mesh see section 3.3.2) and `SOURCES` (local control sources see section 3.3.3).

○ 3.2 **load nodal mesh parameters for the coarse mesh**

If mesh parameter controlled mesh generation is selected then this procedure will load nodal mesh parameters from file into the mesh definition data structure of the coarse mesh. The parameters will be copied into an internal variable: the mesh parameter vector is as declared in section 6.2.5.

○ 3.3 **load background mesh definition**

When the element sizes in the mesh generation operation are controlled by a background mesh with or without local control sources, the background mesh variables described in Section 6.2.5 have to be loaded, checked, parsed and updated into the format of the internal data structures. First the background mesh is read from file into a mesh definition data structure.

○ 3.4 **check all triangles are ACW**

As the triangles of the background mesh will be used for geometrical searching procedures it must be verified that all triangles are defined anti-clockwise.

○ 3.5 **load background mesh parameters**

The mesh parameters attached to the background mesh are loaded from file into the the background mesh definition data structure. Subsequently they are duplicated into the internal format of a mesh parameter vector.

○ 3.6 **setup adjacency for background mesh**

The adjacency related information of the background mesh is determined and stored. This information is required for the searching procedures which will be carried out on the background mesh to determine mesh sizes in the mesh generation procedure. The method to retrieve this information is completely identical to the method applied to retrieve the coarse mesh adjacency data which will be discussed as step 1.3 of Figure 6.6 on page 97.

○ 3.7 **load sources definition**

When local control sources are required for determination of the mesh sizes the definition of these sources, as described in Section 3.3.3, is loaded from a file into a sources definition data structure.

3.4 Concluding remarks

In this chapter the general layout for the mesh generation software has been described. Methods for the control of the mesh sizes generated have been outlined. In the next chapter the first unstructured mesh generation method discussed in this book, the advancing front method will be described. In Chapters 5 and 6, two other unstructured mesh generation techniques will be covered.

Bibliography

[1] Blacker, T.D., Stephenson, M.B., "Paving: A new approach to automated quadrilateral mesh generation", *International Journal for Numerical Methods in Engineering*, 32: 811–847, 1991.

[2] Blacker, T.D., Stephenson, M.B., Canann, S., "Analysis automation with paving: A new quadrilateral meshing technique", *Advances in Engineering Software*, 56: 332–337, 1991.

[3] Cheng, B., *Parallel adaptive finite element mesh generation*, M.Phil. Thesis, Heriot-Watt University, Edinburgh, UK, 2001.

[4] George, P.L., Borouchaki, H., *Delaunay Triangulation and Meshing, Application to Finite Elements*, Editions Hermes, Paris, France, 1998.

[5] Lo, S.H., "A New Mesh Generation Scheme for Arbitrary Planar Domains", *International Journal for Numerical Methods in Engineering*, 21(8): 1403–1426, 1985.

[6] Lohner, R., Parikh, P., "Three-Dimensional Grid Generation by the Advancing Front Method", *International Journal for Numerical Methods in Fluids*, 8(10): 1135–1149, 1988.

[7] Peraire, J., Vahdati, M., Morgan, K., Zienkiewicz, O.C., "Adaptive Remeshing for Compressible Flow Computations", *Journal of Computational Physics*, 72(2): 449–466, 1987.

[8] Shewchuk, J.R., *Delaunay Refinement Mesh Generation*, PhD thesis, Carnegie Mellon University, Pittsburgh, PA, USA, 1997.

Chapter 4

The Advancing Front Method

4.1 Introduction

The advancing front method has proved to be a popular mesh refinement algorithm. It was first presented by Lo in 1985 [9]. He presented a new mesh generation algorithm that initially defines the boundary of the domain by a set of line segments, the initial advancing front. All nodes are generated within the domain bounded by the initial advancing front and valid triangular elements are then formed from the line segments and interior nodes. Each element is generated by joining the end nodes of a line segment to a third node with the condition that it does not intersect the advancing front. The front is then updated and the element creation process goes on as long as the front is not empty.

Peraire *et al* [13] presented a new version of the advancing front method in 1987 for steady state solutions of the Euler equations in two dimensions. Most of the subsequent advancing front method research has been based upon Peraire's algorithm. Their paper [13] describes an algorithm whereby a boundary of line segments is generated in a manner similar to the procedure described by Lo [9]. Triangular elements are generated by finding, or generating, a node for an active line segment on the front, therefore nodes and elements are generated simultaneously. When a new element is generated the boundary of line segments is updated by removing corresponding existing line segments from the front and adding new segments to the front. Refinement continues in this manner until the front is empty. The most significant feature of this new advancing front method is the directionality. This makes it superior to Delaunay triangulation, which will be described in detail in Chapter 6, for problems requiring directionality.

The advancing front algorithm, in three dimensions, was first described by Lohner and Parikh [10]. They describe an algorithm based upon the two-dimensional algorithm presented in reference [13]. The advancing front is formed by triangular faces and valid tetrahedral elements are formed by locating a suitable fourth node for each face on the front. Procedures are described to ensure that two triangular faces in three-dimensional space do not intersect each other. Suitable data structures for the three-dimensional algorithm are also presented which minimise the search overhead incurred by the algorithm.

Peraire *et al* [12] also went on to extend their two-dimensional mesh generation algorithm to three dimensions for finite element Euler computations. Their paper describes the first adaptive mesh generation procedure in three dimensions. The size of the generated elements is controlled by parameters assigned to a background mesh. Jin and Wiberg [7] describe an advancing front algorithm in two dimensions. Their research is principally based upon the work of Peraire *et al* [13] but they also use the data structures introduced by Lohner and Parikh [10]. Adaptive remeshing is undertaken by the inclusion of isolines which specify the sizes of the elements to be generated. This removes the need to construct a background mesh manually.

By the early 1990s the principles of the advancing front method were well established and had been implemented successfully by a number of researchers. A very useful paper presented by Jin and Tanner [6] provided a round survey concerning the robustness of the advancing front method. They describe three common problems encountered with the advancing front method in three dimensions, robustness, quality and processing time. They remark that robustness and quality of the mesh are difficult to guarantee. Furthermore extensive searching, sorting and checking routines are required to generate valid elements and these increase the computational time for mesh generation. Jin and Tanner [6] presented two efficiency tests. The first is a validity test and the second is an auxiliary test which seeks to increase the robustness of the advancing front algorithm in three dimensions. The validity test checks if a potential tetrahedral element intersects the advancing front and the auxiliary test checks that the best element is generated from a face on the front. They subsequently found that the inclusion of the auxiliary test increased the robustness of the algorithm and gave satisfactory results.

A research paper questioning the robustness of the advancing front method was published by Moller and Hansbo [11]. They suggest various other methods of increasing the robustness of the algorithm. For example, if a face on the front cannot be used to generate an element it is placed in a list of failed faces and the procedure continues with another face. In this way it is hoped that failed faces will be incorporated into alternative elements. These procedures have been successful in increasing the robustness of the algorithm. A description of the generation of exceptionally stretched tetrahedral elements adjacent to the boundary is also given. The robustness of the advancing front method can be further increased by incorporating principles of Delaunay triangulation [4] into the algorithm. The advancing front method has been found to be a slow refinement algorithm when compared to Delaunay triangulation [4]. This has generated a large amount of research with the objective of increasing the computational efficiency of the algorithm.

Bonet and Peraire [2] have developed useful data structures to aid the efficiency of the advancing front algorithm. These reduce the number of searches by organising the data into logical structures which greatly increase the success of the algorithm and provide near linear processing times in terms of the number of elements. In 1995, Baciu and Tsang [1] applied the advancing front method to generate meshes for the solution of radiosity rendering problems. The authors emphasise the importance of the computational speed of the generation of acceptable grids. They subsequently present two mesh refinement algorithms based upon the advancing front method which are described as efficient. Computational time for refinement has also been decreased by Kwok *et al* [8] through the development of efficient data structures for searching, inserting and deleting sides during the algorithm. Efficiency of the order $O(\log N)$ has been reported.

In 1996 Hassan *et al* [5] presented a paper on the advancing front method by concentrating on the generation of stretched elements for the analysis of viscous flow simulations. The numerical performance of the algorithm was demonstrated and further research was constrained by the availability of significant computational resources and the improvement of the performance of flow solvers.

4.1.1 Parallel advancing front mesh generation

Implementation of a parallel algorithm for the advancing front method was presented by Topping and Khan [14]. Their starting point is a background mesh of elements which is decomposed into a number of subdomains. Refinement proceeds by generating an advancing front for each background element and refining the elements within each subdomain individually. Parallelisation is therefore based upon a static explicit division of the domain. By using nodal adaptive parameters and ordering the advancing front for boundary faces, boundary conformity between adjacent subdomains is achieved with no inter-processor communication. Efficient domain decomposition programs would subsequently be required to ensure load balancing amongst the worker tasks and to minimise the number of nodes on the boundary of two or more subdomains. This parallel work was extended to three dimensions by Wilson and Topping [15].

4.2 Overview of the advancing front algorithm

The layout of the implementation of the advancing front method, as implemented by the authors is shown in Figure 4.1. The main features of the algorithm are as follows:

- 1. **Input**
  The program loads in a coarse mesh, determines its boundary and refines each boundary segment individually. This refined boundary is the initial front for the method.

- 2. **More than three segments**
  If there are only three segments on the front further processing is stopped and the three segments form one final triangle.

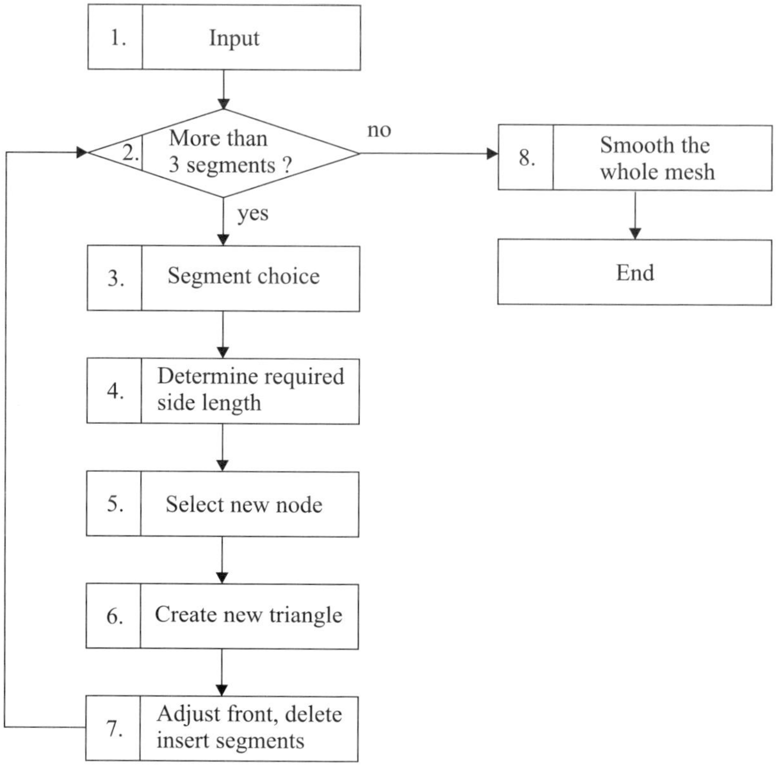

Figure 4.1: Overview of the advancing front algorithm.

◦ 3. **Segment selection**
When there are more than three segments on the front, the reference or base segment is selected. The segment of shortest length is selected as a base segment which will also serve as one side of the new triangle to be formed.

◦ 4. **Determine required side length**
The length of the sides of the new triangle element is based on the specified element size values at the ends of the base segment.

◦ 5. **Select new node**
By checking already existing nodes and some newly created nodes, the "best" possible node is selected. Best, in the sense that the triangle formed by the end nodes of the base segment and the selected node satisfies the side length criterion, the triangle has a good shape and it does not intersect any other segments of the front.

- 6. **Create new triangle**
  When a node is successfully selected a new triangle element is formed.
- 7. **Adjust front, delete, insert segments**
  After the element creation the current front is adjusted. The base segment is removed from the active front while the two other sides of the new triangle (considering them individually) are either added to the active front (if they were not already on it) or they are removed (if they are already on the active front).
- 8. **Smooth the whole mesh**
  When all the elements have been created, the mesh is smoothed and adjusted to achieve better quality elements.

In the next sections the details of these stages are discussed in further detail.

4.3 Input data

The input data for the advancing front method is a list of boundary segments. The external boundary should be defined in an anti-clockwise manner while all internal loops should be specified in a clockwise manner. The boundary is refined according to the mesh parameters given at the meshpoints by the methods given in Appendix A. The mesh parameter at a node can be determined by one of the three methods described in Chapter 3.

The list of refined boundary segments is used as the initial front for the method. The order of the segments is not important. It can be arbitrary. However the segments are directed and an imaginary vector points from the first node to the second node of each segment.

It should be noted that, as a result of the format of the input data, it is also possible to refine a coarse mesh on an element-by-element basis (see Section 1.3.5). In this case each element in the coarse mesh is considered in a sequence. Special care must be taken with boundary segment generation to avoid duplicate nodes. If all edges are refined only once and the refined boundary is reused when required the problem of duplicate nodes can be avoided. This procedure ensures the compatibility between sides where the two elements meet. The method of providing the refined boundary segments of an element to the mesh generator refines the internal parts of the coarse element. When all coarse elements are refined internally no further processing (such as double node removal) is necessary and the mesh can be assembled.

4.4 Base segment selection

When the method selects the base segment it only considers the length of the segment. The shortest segment on the current front is selected as the base segment. The reason for this selection criterion is that in the case of adaptive meshing the transition from small to large elements occurs further away from the small elements ensuring a more robust mesh generation method. The selection of the shortest segment also ensures that the generation of large elements does not crowd out small elements which have yet to be generated.

4.5 Required side length determination

The required side length or mesh parameter (δ) is always calculated at the middle of the base segment. Using the coarse mesh with the mesh parameters provided, a quadratic function can generally be used to determine the mesh parameters

$$d = \frac{\sum_{i=1}^{n} \left(\frac{1}{||X - X_i||} \right)^2 d_i}{\sum_{i=1}^{n} \left(\frac{1}{||X - X_i||} \right)^2} \tag{4.1}$$

where the d is the calculated mesh parameter at position X, d_i is the mesh parameter at position X_i of a meshpoint i and n is the number of boundary meshpoints including the boundary meshpoints in the coarse mesh and the refined boundary meshpoints as well. The function ensures a smooth transition between largely varying mesh parameters. The use of the function as a sizing function was suggested by Class *et al* [3].

The other two methods of determining the mesh parameters as discussed in section 3.3 have a limited use when applied to the advancing front method. The reason is that the method has to start from the smallest segment to ensure the robustness of the algorithm. Therefore when a local mesh parameter source is defined inside the background mesh with a smaller mesh parameter than in the background mesh the algorithm cannot start from the smallest segment, since that segment is somewhere inside the mesh and the algorithm starts from the boundary. This means that only local control sources with larger mesh parameters than in the background mesh should be specified for the advancing method. In this way, a fine boundary and coarse internal meshes can be generated.

4.6 Node selection algorithm

The node selection algorithm is shown in Figure 4.2. The algorithm starts by generating a series of chord nodes which are perpendicular to the base segment as shown in Figure 4.3. (In the authors' implementation the number of generated chord nodes can be controlled by a compilation flag which typically has the value of five.) The distance d is determined in such a way that the length of the sides will be equal to the required length δ.

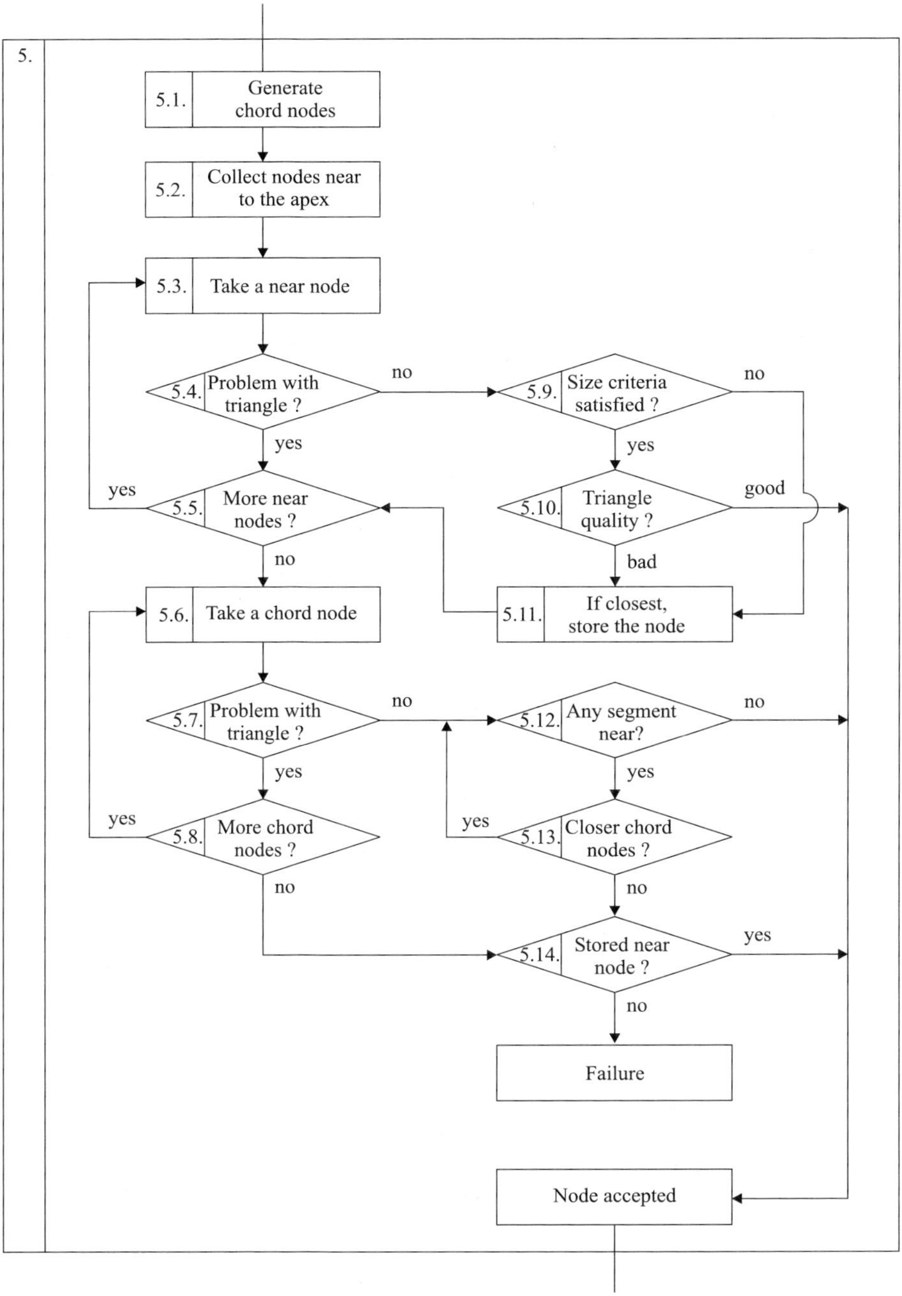

Figure 4.2: Node selection algorithm.

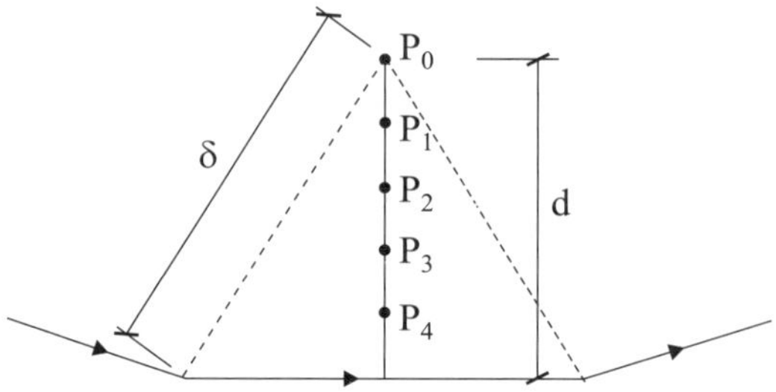

Figure 4.3: Generation of chord nodes.

In the next step the nodes in the proximity of the apex of the chord node (node P_0 in Figure 4.3) are found. Only a predetermined number of these 'near' nodes is stored to speed-up the algorithm. (The number of near nodes can be controlled by a compilation flag. Typically a limit of 20 is set on the number of near nodes.) The distance is measured between the apex chord node and all candidate nodes. Only those nodes on the current front which are on the 'left hand side' of the segment are considered. This criterion eliminates all nodes with which the triangular elements formed would be defined in a clockwise manner. For example in Figure 4.4 if $\overline{ab}$ is the base segment then node d would not be acceptable as a near node. Not only would the triangle formed be outside the domain to be meshed, but the triangle orientation would be clockwise. However node c would be an acceptable near node. All nodes falling in the shaded area (on the left hand side of segment $\overline{ab}$) are acceptable near nodes. The collected near nodes are sorted according to their distance to the apex chord node.

The algorithm first examines all near nodes and checks that sides of the triangle formed by the particular near node and the end nodes of the base segment do not intersect any other front segments. It also checks that no other segment of the front would fall inside the triangle formed. When any of these problems are encountered the near node is ignored. A further criterion is that the length of the sides of the triangle formed should be smaller than or equal to the required length multiplied by a specified factor. The factor larger than 1.0 allows sides longer than the required length to be accepted. This approach favours existing nodes thus avoiding, when possible, new node generation. Finally the quality of the formed triangle is assessed using

$$F_i = 1.0 - \frac{\sum_{j=1}^{3} e_j \; e_{((j+1) \; mod \; 3)}}{3} \tag{4.2}$$

where F_i is the value of quality for element i, e_j is the unit vector of edge j of element i and $mod3$ is the modulo division by three. In this context the result of a modulo division is the remainder of an integer division (for example $10 \; mod \; 3 = 1$). The minimum value of Equation 4.2 is 0.0 when two internal angles are zero degrees and one angle is 180 degrees. The higher the value the better the quality of the triangle. For an equal sided triangle F_i is equal to 0.5.

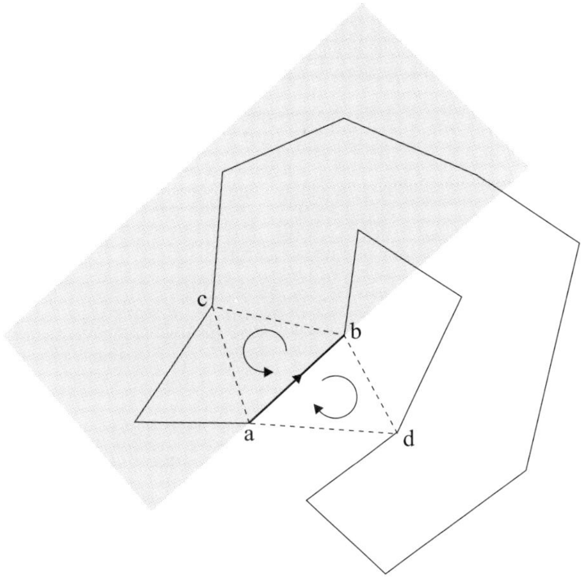

Figure 4.4: Near node selection.

If all criteria are met then the node is immediately accepted as the third node of the triangle. If the triangle does not intersect any other segment but does not satisfy the other two criteria the node is stored for later use.

If no satisfactory near node can be found the algorithm tries to use the chord nodes. It starts from the apex chord node and considers each node towards the base segment. If the formed triangle (the base segment with the particular chord node) does not intersect any other front segment, the node is accepted as a candidate. A second test for the candidate chord node was suggested by Topping and Khan [14] where the candidate is examined to determine if it is too close to a segment of the front. A front segment is considered to be close to the particular chord node in two cases:

- when a line perpendicular to the examined front segment intersects the examined segment between its two end points and the distance between the chord point and the front segment is smaller than half of the distance between the chord node and the middle of the base segment (see Figure 4.5(a), where $d < c_d/2$) or

- one of the end nodes of the front segment is closer to the chord node than half of the distance between the chord node and the middle of the base segment (see Figure 4.5(b), where $d < c_d/2$).

These conditions are to ensure that the current triangle element and the element to be generated from the other segment which is close are equally spaced. An example is shown in Figure 4.6.

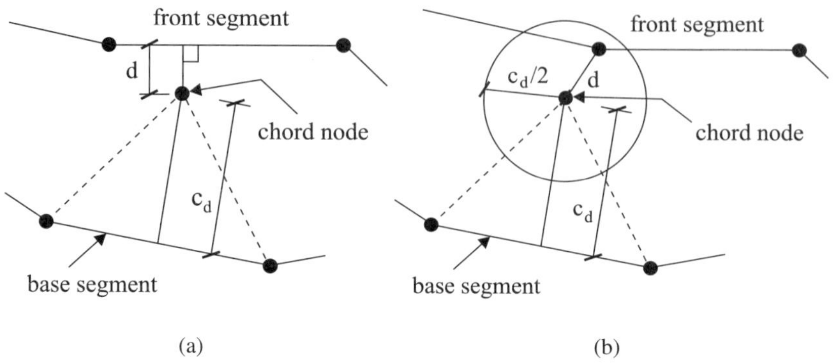

Figure 4.5: Definition of closeness.

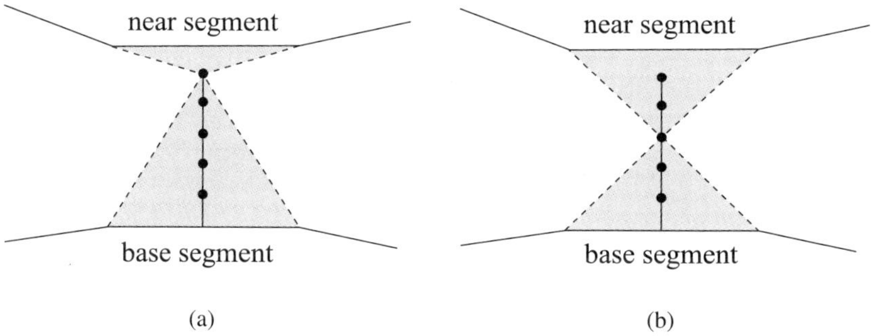

Figure 4.6: Triangle element generation (a) when not considering 'near' segments and (b) when 'near' segments are taken into account.

If a segment is so close that none of the chord nodes could be accepted then the algorithm tries to revert back to a stored near node. This assumes that the stored node may be an end node of the particular close front segment. If there was no near node stored then the algorithm fails.

4.7 Element creation and front adjustments

After the node is selected a new element can be created. The algorithm ensures that all elements will be defined in an anti-clockwise manner.

When the element is created some adjustments have to be made to the active front.

The base segment has to be removed from the front. In the case of the other two sides of the triangle the adjustment depends on whether the particular segment exists on the front or not. If the segment does exist on the front it means that the new element has been generated using the segment, therefore it should be removed from the front. On the other hand when the segment does not exist before the element creation it should be added to the active front. Because the order of the segments on the front does not matter, the new segments are simply added to the end of the list of the segments on the active front.

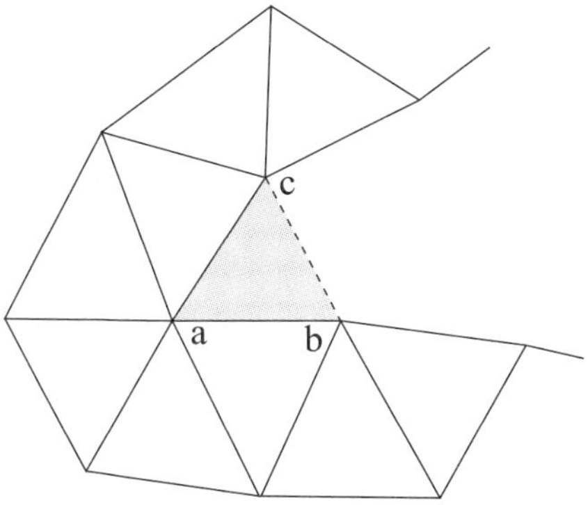

Figure 4.7: Active front adjustment.

Figure 4.7 shows an example, where the base segment is $\overline{ab}$ and the selected node is c. The new triangle is the shaded element. It can be seen that not only segment $\overline{ab}$ should be removed but the other existing side segment $\overline{ac}$ as well. However segment $\overline{bc}$ becomes a new segment on the active front.

4.8 Concluding remarks

In this chapter the advancing front method has been described. The implementation issues have been detailed. Examples of meshes generated using the advancing front method have already been shown in Chapter 1. Further examples will be given in Chapter 10. In Chapter 8, techniques will be discussed that may be used to convert triangular meshes generated from the advancing front technique into quadrilateral meshes.

In the next chapter the paving method for generating quadrilateral meshes directly without the need to generate triangular meshes is discussed. The paving method may be considered as another version of the advancing front method, in that the objective of the method is to mesh the interior of the domain from a gradually reducing boundary.

Bibliography

[1] Baciu, G., Tsang, R.K.W., "Advancing Front Meshing for Radiosity Solutions", *Lecture Notes in Computer Science*, 1024: 283–291, 1995.

[2] Bonet, J., Peraire, J., "An Alternating Digital Tree (ADT) Algorithm for 3D Geometric Searching and Intersection Problems", *International Journal for Numerical Methods in Engineering*, 31(1): 1–17, 1991.

[3] Cass, R.J., Benzley, S.E., Meyers, R.J., Blacker, T.D., "Generalized 3-D paving: An automated quadrilateral surface mesh generation algorithm", *International Journal for Numerical Methods in Engineering*, 39: 1475–1489, 1996.

[4] George, P.L., Borouchaki, H., *Delaunay Triangulation and Meshing, Application to Finite Elements*, Editions Hermes, Paris, France, 1998.

[5] Hassan, O., Morgan, K., Probert, E.J., Peraire, J., "Unstructured Tetrahedral Mesh Generation for Three-dimensional Flows", *International Journal for Numerical Methods in Engineering*, 39(4): 549–567, 1996.

[6] Jin, H., Tanner, R.L., "Generation of Unstructured Tetrahedral Meshes by Advancing Front", *International Journal for Numerical Methods in Engineering*, 36(11): 1805–1823, 1993.

[7] Jin, W., Wiberg, N.E., "Two-dimensional Mesh Generation, Adaptive Remeshing and Refinement", *International Journal for Numerical Methods in Engineering*, 29(7): 1501–1526, 1990.

[8] Kwok, W., Haghighi, K., Kang, E., "An Efficient Data Structure for the Advancing Front Triangular Mesh", *Communications in Numerical Methods in Engineering*, 11(5): 465–473, 1995.

[9] Lo, S.H., "A New Mesh Generation Scheme for Arbitrary Planar Domains", *International Journal for Numerical Methods in Engineering*, 21(8): 1403–1426, 1985.

[10] Lohner, R., Parikh, P., "Three-Dimensional Grid Generation by the Advancing Front Method", *International Journal for Numerical Methods in Fluids*, 8(10): 1135–1149, 1988.

[11] Moller, P., Hansbo, P., "On Advancing Front Mesh Generation in Three Dimensions", *International Journal for Numerical Methods in Engineering*, 38(21): 3551–3569, 1995.

[12] Peraire, J., Peiro, J., Formaggia, L., Morgan, K., Zienkiewicz, O.C., "Finite Element Euler Computations in Three Dimensions", *International Journal for Numerical Methods in Fluids*, 26(10): 2135–2159, 1988.

[13] Peraire, J., Vahdati, M., Morgan, K., Zienkiewicz, O.C., "Adaptive Remeshing for Compressible Flow Computations", *Journal of Computational Physics*, 72 (2): 449–466, 1987.

[14] Topping, B.H.V., Khan, A.I., *Parallel Finite Element Computations*, Saxe-Coburg Publications, Edinburgh, UK, 1996.

[15] Wilson, J.K., Topping, B.H.V., "Parallel adaptive tetrahedral mesh generation by the advancing front technique", *Computers & Structures*, 68(1-3): 57–78, 1998.

Chapter 5

The Paving Method

5.1 Introduction

The paving method was initially proposed, in 1990, by Blacker and Stephenson [5–7] and the first implementation was done at Sandia National Laboratories [4].

5.1.1 Main features

The main features of the paving method are:

- The method is an advancing front style mesh generation, where by adding elements layer-by-layer to the boundary the mesh generation propagates towards the internal parts of the mesh.
- The method generates an all quadrilateral mesh.
- The mesh generation is independent of the orientation of the coarse mesh and it follows the features of the boundary.
- The method not only smoothes the mesh when the mesh generation is finished but it uses an incremental smoothing technique during the mesh generation.

The paving method is patented [2] but several researchers have utilised and modified it in their research projects. The method has also been used on distributed computers by Lober *et al* [16].

5.1.2 Paving in three-dimensions

Cass *et al* [8] generalised the paving method for three dimensions and White and Kinney [27] redesigned it by changing the row-based approach to an element-by-element approach. The resulting paving method is more robust and was implemented in Sandia

National Laboratory [3, 21]. The three dimensional equivalent of paving is plastering which was developed by Stephenson *et al* [24].

5.1.3 Applications and the paving method

Kawamura *et al* [12] used the method in the simulation of crack propagation. They also proposed some changes to the paving method to account for the large differences in the size of small elements around a crack and the larger surrounding elements. The modifications mainly change the smoothing process. Ladeinde and Alabbi [15] introduced a "set of universal rules" for the front closure problem in the paving method which are "applicable throughout the domain" and do " not require special consideration at closure regions, concave or convex". Jang *et al* [11] developed a surface mesh generator using the paving method to simulate the forming of curved plates by line heating. Owen *et al* [18] developed a groundwater modelling system which can also use the paving method. Nowottny [17] and Aleksa *et al* [1] also used paving, but in their approach the domain to be meshed is decomposed into smaller, "simple" areas geometrically and these simple areas are paved individually. In this way the time spent in meshing can be reduced. Owen *et al* [19] proposed a method to overcome the difficulties of the paving method and provide a more robust all quadrilateral mesh generation method. Their method, called Q-Morph, uses the principle of the paving method to transform triangular elements into quadrilateral elements in an advancing front fashion. In another approach Rees [20] developed a mesh generation method which is also based on paving but among the quadrilateral elements the method is allowed to generate triangle elements to avoid distorted quadrilateral elements. The paving method can also be used to generate triangular elements with directional control as was demonstrated by van Rens [25, 26].

5.2 Overview of the paving method

The layout of the implementation of the paving process is shown in Figure 5.1.

- 1. **Input**
  The program loads a coarse mesh, determines its boundary and refines each boundary segment individually. The longest segment is refined last in such a way as to ensure that finally the mesh boundary will contain an even number of nodes on the mesh boundary. In Section 5.3 further details concerning the input will be presented.

- 2. **Row selection**
  At this step the algorithm attempts to determine where to start the element generation on the paving loop. The first element in the row is also generated at this stage. Further details are given in Section 5.5.1 and 5.5.2.

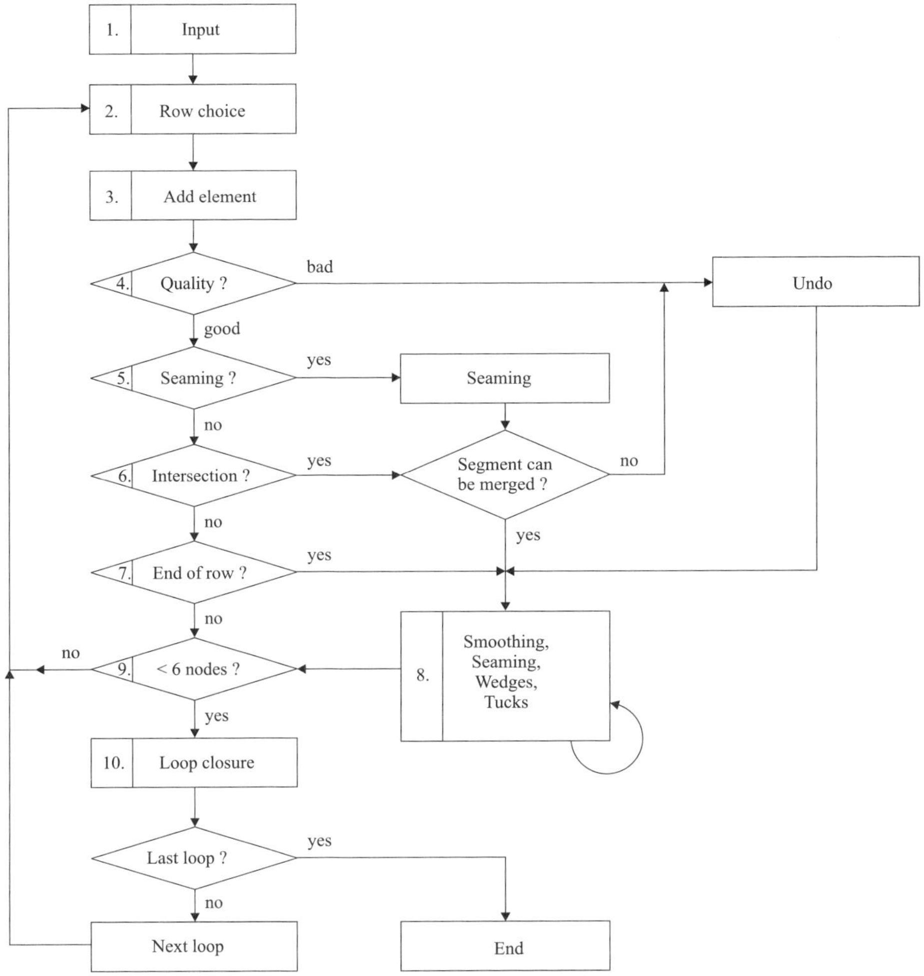

Figure 5.1: Overview of the paving algorithm.

- 3. **Element addition**
  A new quadrilateral element is added to the mesh at this stage according to the rules in Section 5.4.5 and Section 5.5.4.

- 4. **Quality check**
  The validity of every generated quadrilateral element is verified. For example an angle larger than 180 degrees would invalidate a quadrilateral element. Quality checks are described in Section 5.4.2.

- 5. **Seaming check**
  When a new element is generated a seaming check is performed to determine

if it is possible to perform seaming at any of the nodes of the new element. Seaming is described in Section 5.4.6.

- ∘ 6. **Intersection check**
  Furthermore every new element is checked for intersection with every paving front and all of the new edges of the new element are checked for proximity with other edges on any of the paving fronts. When an intersection or proximity is detected the algorithm attempts to resolve this by merging two segments. In this way paving fronts can be merged or cut into two loops. If the intersection cannot be resolved, because for example the generated quadrilateral elements would be invalid or of bad quality, then the last intersecting element is deleted. Intersection and proximity checks are described in Section 5.6.

- ∘ 7. **End of row**
  At this step the algorithm checks whether the generated row is complete when the generation stops and other operations will be performed on the paving front.

- ∘ 8. **Smoothing, seaming, wedges and tucks**
  When the element generation has been finished on the current paving front or it has been interrupted by either segment merging or seaming, then several adjustments are made to the current paving front. First the row is smoothed and, if necessary, small angles are removed by seaming. Finally, if necessary, wedges are inserted or tucks are formed to preserve the required element sizes. If seaming, wedge insertion or tuck formation is performed then another smoothing step is carried out. This procedure is repeated until smoothing is the only operation being performed. Smoothing is described in Section 5.7, seaming is explained in Section 5.4.6 while wedge insertion and tuck formation is presented in Section 5.4.7.

- ∘ 9. **Less than six nodes**
  If the number of nodes on the current paving row is fewer than six, special loop closure procedures are applied.

- ∘ 10. **Loop closure**
  Paving fronts with fewer than six nodes are closed by specially formulated procedures as described in Section 5.8. If the closed loop was the last loop in the mesh, then the algorithm quits.

5.3 Input data

The input data to the paving method are closed loops of ordered nodes. An example is shown in Figure 5.2. There can be only one *external loop* and several *internal loops* inside the external one. The loops should not intersect or touch each other. The nodes should be defined in an anti-clockwise order on the external loop, and in a clockwise order on the internal loops. None of the nodes on these input loops can be deleted

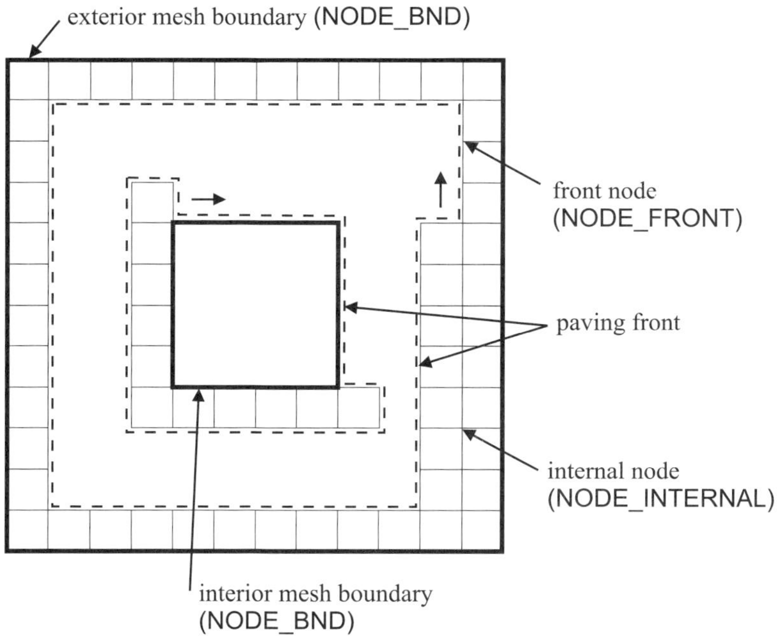

Figure 5.2: Mesh boundaries, paving fronts and node types.

and their coordinates cannot be changed. Furthermore no additional nodes can be inserted into these loops. These mesh boundary nodes are marked as NODE_BND in the program. The number of nodes on each loop must be even to ensure that an all quadrilateral mesh can be generated.

Initially the *paving fronts* are the input mesh boundary loops and, as new nodes and elements are generated, the paving front moves away from the mesh boundary. There is also only one external paving front at the beginning, but as the paving fronts are merged or cut several external paving fronts may form. The nodes on the paving fronts are denoted in the implementation presented here as NODE_FRONT and the interior nodes as NODE_INTERNAL. These node types are also shown in Figure 5.2.

5.3.1 Implementation details of input data

The actual input data to the program is a coarse mesh with mesh parameters. (For the discussion about mesh parameters see Chapter 3.) The boundary segments of the input coarse mesh are extracted and refined according to the specified mesh parameters. This procedure will also be used in the two-dimensional Delaunay mesh generation described in section 6.5.

To help the refinement of the segments, there is no restriction on the number of generated nodes in the segments, except for the last segment. The longest boundary segment of the extracted boundary is refined last, and the number of generated nodes on this segment is generated in such a way that the total number of nodes on the whole boundary loop will be even. In this way the refinement method is allowed to satisfy the mesh parameters as much as possible, and the small intervals in the refinement are less likely to be crowded out. The segment refinement procedure is discussed in Appendix A.

The background meshes and the local control sources in the calculation of the mesh parameters have limited use as a result of the same problems with the advancing front method as discussed in Section 4.5.

5.4 Foundations of the paving method

This section contains some definitions and basic operations which are used later to describe the implementation of the algorithm in greater detail.

5.4.1 Element definition

A valid element is defined by four nodes in an anti-clockwise manner where none of the internal angles is less than 10 degrees or larger than 180 degrees. All quadrilateral elements shown in Figure 5.3 are invalid.

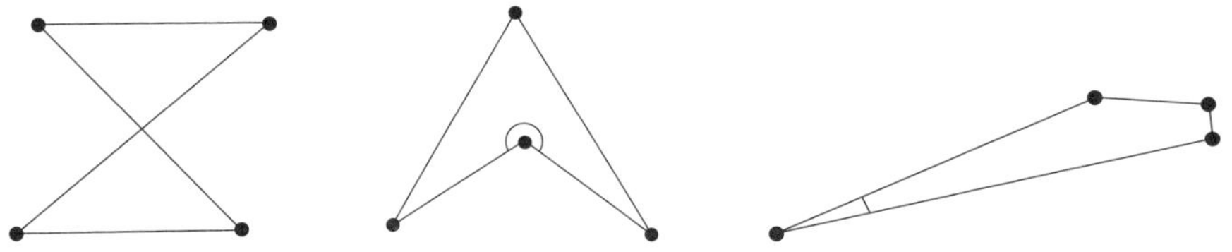

Figure 5.3: Invalid quadrilateral elements, (a) flipped shape, (b) arrowhead shape, (c) interior angle $<$ 10 degrees.

5.4.2 Element quality

If a quadrilateral element is valid, as described above, the quality of the element is determined using

$$F_i = 1.0 - \frac{\sum_{j=1}^{4} |e_j \; e_{((j+1) \; mod \; 4)}|}{4} \tag{5.1}$$

where F_i is the value of quality for element i, e_j is the unit vector of edge j in element i, $mod\ 4$ denotes a modulo division by four and the vertical lines denote the absolute value. The value of the function is between 0.0 and 1.0 where the value of 0.0 is deemed unacceptable and the value of 1.0 represents a quadrilateral element with "perfect" perpendicular edges.

Furthermore the length of every side is compared with the length of the adjacent sides. If the ratio of the lengths is greater then 2.0 then the element is considered to be of bad quality. Though this quality measure is not dominant. If this second condition is not satisfied, the element may be still accepted.

5.4.3 Element configuration

During the paving process the preferable configuration of elements to be generated is shown in Figure 5.4(a). In Figure 5.4 the marked node is part of four quadrilateral elements and is called a *regular* node. In all other cases when a node is part of fewer or more than four quadrilateral elements the node is called an *irregular* node. Some examples for irregular nodes are shown in Figure 5.4(b).

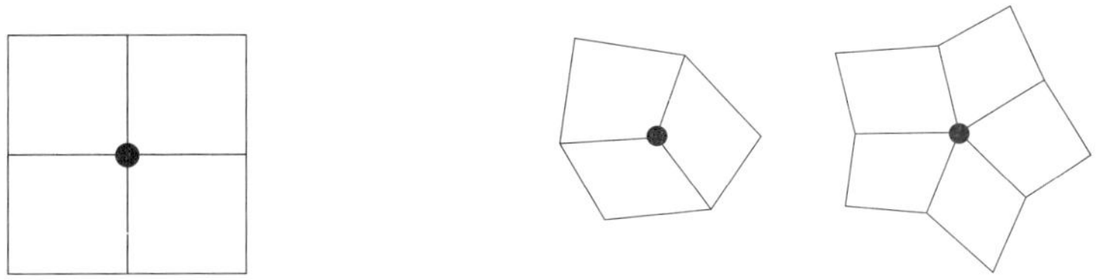

(a) (b)

Figure 5.4: (a) Regular and (b) irregular node configuration.

5.4.4 Node classification

Each node on the current paving front is classified according to the internal angle at the node. The internal angle, α as shown in Figure 5.5, is defined as the angle between two vectors, $\mathbf{a}$ and $\mathbf{b}$. Vector $\mathbf{a}$ points from the previous node $(i-1)$ to the current node (i) and vector $\mathbf{b}$ points from the current node (i) to the next node $(i+1)$ on the front. The dot product of the two vectors provides the angle γ:

$$\cos\gamma = \frac{\mathbf{a}\,\mathbf{b}}{|\mathbf{a}|\,|\mathbf{b}|} \tag{5.2}$$

The cross product of the vectors $\mathbf{a}$ and $\mathbf{b}$ is a vector of $\{0\ 0\ z\}$ since the program works in the $x-y$ plane. Moreover considering that the external boundary front is

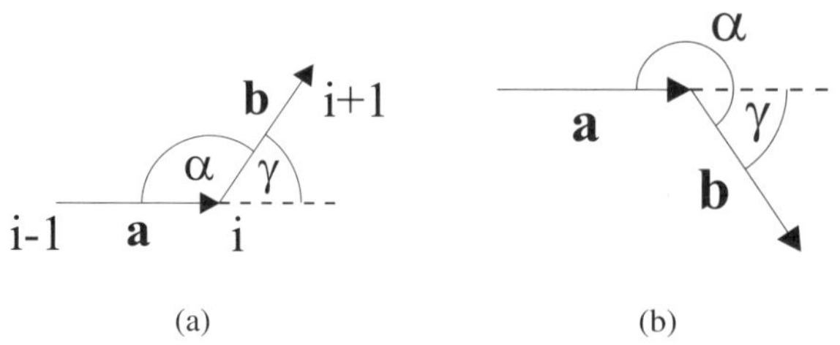

Figure 5.5: Internal angle calculation.

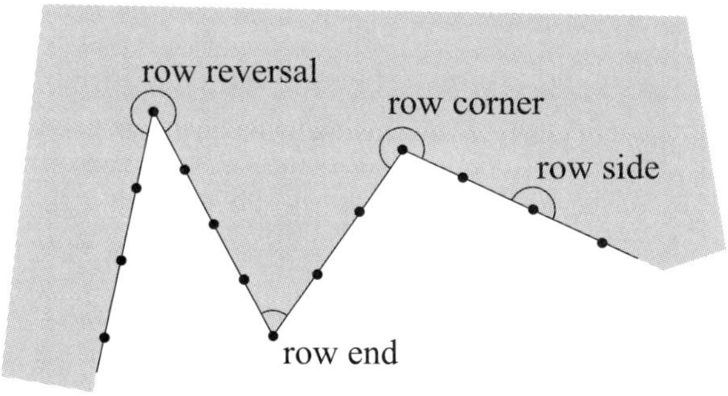

Figure 5.6: Node types and internal angles for an example boundary.

anti-clockwise oriented and the internal boundary fronts are clockwise oriented the internal angle, α, is:

$$\alpha = \left\{ \begin{array}{ll} \pi - \gamma & \text{if } z \geq 0 \\ \pi + \gamma & \text{otherwise} \end{array} \right\} \tag{5.3}$$

Some interior angles are also shown in Figure 5.6.

During the classification the calculated internal angle is compared with a series of angle ranges subject to predefined angle tolerances ($\sigma_1, \ldots \sigma_6$). These tolerances allow some deviations from the expected angle range, as shown in Figure 5.7. The values of the tolerances in the implementation presented here were determined by trial and error. Considering the internal angle (α) with the tolerances ($\sigma_1, \ldots \sigma_6$) the following types of nodes are possible:

- **row end**: when $\alpha \leq \pi/2 + \sigma_1$
- **ambiguous row end or side**: when $\pi/2 + \sigma_1 < \alpha \leq \pi - \sigma_2$
- **row side**: when $\pi - \sigma_2 < \alpha \leq \pi + \sigma_3$

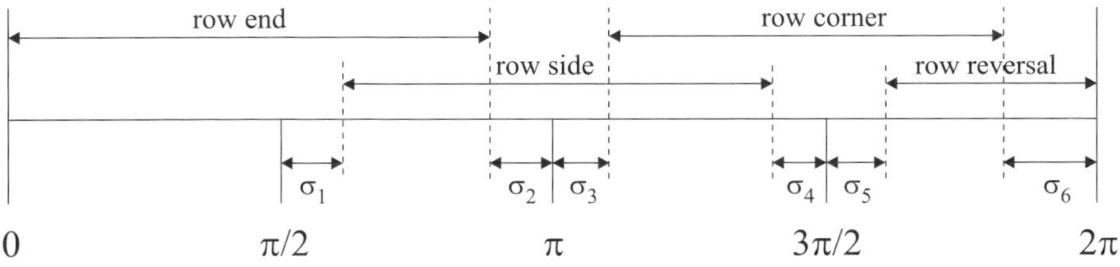

Figure 5.7: Definition of node types and angles as shown in Figure 5.6.

- **ambiguous row side or corner**: when $\pi + \sigma_3 < \alpha \leq 3\pi/2 - \sigma_4$
- **row corner**: when $3\pi/2 - \sigma_4 < \alpha \leq 3\pi/2 + \sigma_5$
- **ambiguous row corner or reversal** : when $3\pi/2 + \sigma_5 < \alpha \leq 2\pi - \sigma_6$
- **row reversal**: when $2\pi - \sigma_6 < \alpha$

An ambiguous node can be either of the two types. It is ensured explicitly in the implementation by the choice of the tolerances that a node can have a maximum of two types. Typical values used by the authors are $\sigma_1 = \pi/4.5$, $\sigma_2 = \pi/6.0$, $\sigma_3 = \pi/4.0$, $\sigma_4 = \pi/4.0$, $\sigma_5 = \pi/18.0$ and $\sigma_6 = \pi/2.5$.

5.4.5 Node and element generation

In Figures 5.8 to 5.11 the creation of new nodes and new elements on the current paving front are shown. The current node is N_i with the internal angle α. It lies on the active front, represented in the figures by a strong black line. New elements are coloured grey and the new parts of the front are shown with a dashed line. The generation procedure is according to the following classification types:

5.4.5.1 Row end node

At a row end node a new node does not have to be generated since a new generated node would be equal to or very close to node N_{i+1} (as shown in Figure 5.8). In this case only one new element is formed by nodes $N_i, N_{i+1}, N_j, N_{i-1}$.

5.4.5.2 Row side node

The generation of a new node at a row side node is as shown in Figure 5.9. On the current paving front node N_i is used as a base point for the generation of the new row side node N_j. The vector from node N_i to N_j is denoted as vector V and it bisects

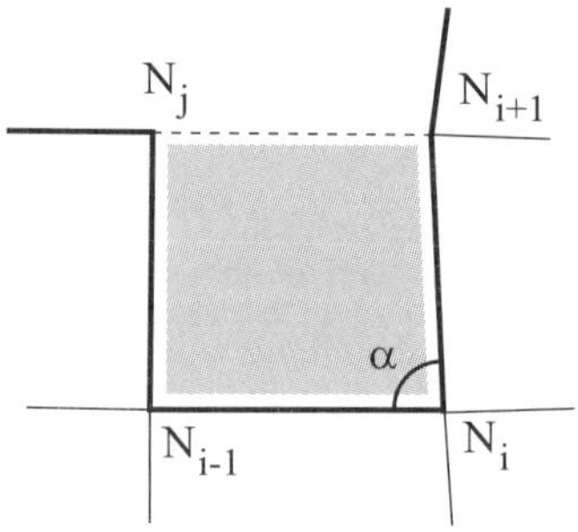

Figure 5.8: Element generation at a row end node.

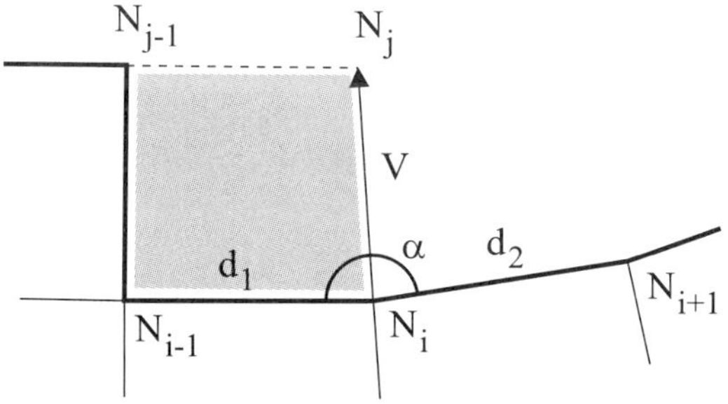

Figure 5.9: Node and element generation at a row side node.

the internal angle (α) at node N_i. The length of the vector V ($|V|$) can be determined using the following formula

$$|V| = \frac{(d_1 + d_2)/2}{sin(\alpha/2)} \tag{5.4}$$

where d_1 is the distance between N_i and N_{i-1} and d_2 is the distance between N_i and N_{i+1}. According to Blacker and Stephenson [6] the sin term in the division should help the generation of a smooth row. The new quadrilateral element is defined by nodes $N_i, N_j, N_{j-1}, N_{i-1}$.

5.4.5.3 Row corner node

The generation of new node at a row corner node is shown in Figure 5.10. On the current paving front node N_i is used as a base point for the generation of the new row corner nodes N_j, N_k and N_l. The vectors from node N_i to N_j, N_k and N_l are denoted as vector V_j, V_k and V_l and they divide the internal angle α at node N_i into 1/3, 1/2 and 2/3 portions in the anti-clockwise manner. The length of the vectors V_j, V_k and V_l

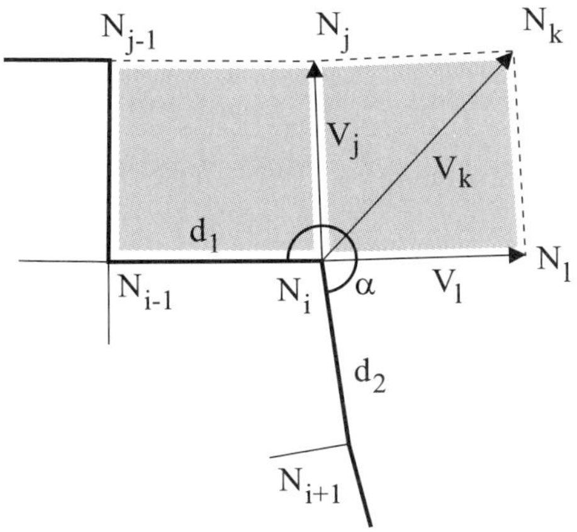

Figure 5.10: Node and element generation at a row corner node.

are determined using the following formulae

$$|V_j| = \frac{(d_1 + d_2)/2}{sin(\alpha/3)} \quad (5.5)$$

$$|V_k| = \sqrt{2}|V_j| \quad (5.6)$$

$$|V_l| = |V_j| \quad (5.7)$$

where d_1 is the distance between N_i and N_{i-1} and d_2 is the distance between N_i and N_{i+1}. In this case two new elements are generated: $N_i, N_j, N_{j-1}, N_{i-1}$ and N_i, N_l, N_k, N_j.

5.4.5.4 Row reversal node

The generation of new nodes at a row reversal node is as shown in Figure 5.11. On the current paving front node N_i is used as a base point for the generation of the new row reversal nodes N_j, N_k, N_l, N_m and N_n. The vectors from node N_i to N_j, N_k, N_l, N_m and N_n are denoted as vector V_j, V_k, V_l, V_m and V_n and they divide the internal angle α at node N_i into 1/4, 3/8, 1/2, 5/8 and 3/4 portions in the anti-clockwise direction. The lengths of the vectors V_j, V_k, V_l, V_m and V_n are determined by the following formulae

$$|V_j| = \frac{(d_1 + d_2)/2}{sin(\alpha/4)} \quad (5.8)$$

$$|V_k| = \sqrt{2}|V_j| \quad (5.9)$$

$$|V_l| = |V_j| \quad (5.10)$$

$$|V_m| = |V_k| \quad (5.11)$$

$$|V_n| = |V_j| \quad (5.12)$$

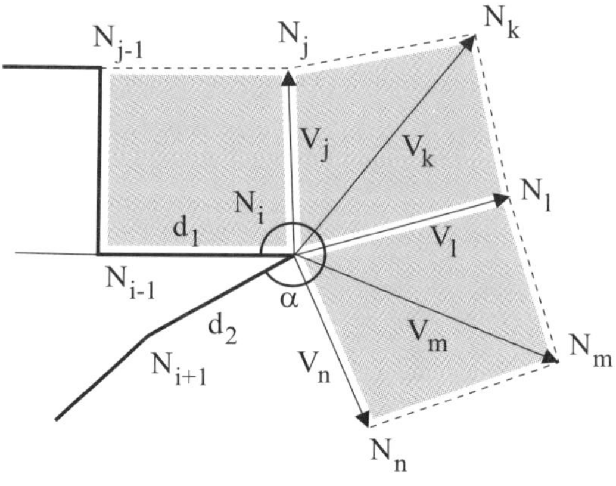

Figure 5.11: Node and element generation at a row reversal node.

where d_1 is the distance between N_i and N_{i-1} and d_2 is the distance between N_i and N_{i+1}. In this case three new elements are generated: $N_i, N_j, N_{j-1}, N_{i-1}, N_i, N_l, N_k, N_j$ and N_i, N_n, N_m, N_l.

5.4.5.5 Element size control

Using the method described above for the element generation, the size of the elements is controlled by the propagating paving boundary. In this case the size of the internal elements is independent of the external size control measures. The only control over the size is provided by the length of the mesh boundary segments.

To enhance the control, as suggested by Cass *et al* [8], a quadratic sizing function can be introduced. The sizing function defines a scalar field which should cover the whole domain. The function is always of a form that can be sampled at any arbitrary point within the domain to be meshed. Using this function the transition from large elements to smaller ones should be smoother. The suggested quadratic sizing function is

$$d = \frac{\sum_{i=1}^{n} \left(\frac{1}{||X - X_i||} \right)^2 d_i}{\sum_{i=1}^{n} \left(\frac{1}{||X - X_i||} \right)^2} \tag{5.13}$$

where the d is the desired size at position X, d_i is the desired size at position X_i of a mesh boundary node i and n is the total number of boundary points. The desired size at a mesh boundary node is determined as the average of the adjacent edge sizes. The calculation of this function can be computationally very expensive, therefore the value of the function is calculated only once when the node is created. It is then stored with the node and assumed not to change significantly with any movement of the node.

Naturally the combination of the two size control methods is possible and other hybrid functions may also be used [8].

5.4.6 Seaming

The seaming process helps to eliminate "small" and "negative" angles from the paving front. An example for a "negative" angle is shown in Figure 5.12(a) where two quadrilateral elements overlap at node N_i. The criteria for seaming depends on the number of elements connected to the node and the angle values. In the case of a "negative" angle the seaming is always performed regardless of other parameters. In the original paper by Blacker and Stephenson [6] three kinds of seaming were identified:

- interior node seaming,
- seaming between elements with unequal side lengths (*transition seaming*) and
- seaming with mesh boundary nodes.

These types of seaming will now be reviewed.

5.4.6.1 Interior node seaming

The interior node seaming is the most commonly performed seaming process during the paving procedure. The conditions to perform this seaming process at a node are

$$\left\{ \begin{array}{c} \alpha < \varepsilon_1 \text{ for } N_E \geq 4 \\ \alpha < \varepsilon_2 \text{ otherwise} \end{array} \right\} \text{ where } \varepsilon_1 > \varepsilon_2 \tag{5.14}$$

where α is the interior angle at the node and N_E is the number of elements connected to the node. This condition was proposed by Kim [13] and it is slightly different from the condition suggested by Blacker and Stephenson [6]. It generally enforces the rule that it is more favourable to seam at a regular node or at a node with more than four elements than at a node with less than four elements. The threshold value ε_1 is set to 35 degrees and ε_2 is set to 10 degrees.

In Figure 5.12 an example of the interior node seaming is shown. Three elements (N_E = 3) are attached to node N_i and the angle is "negative" which means that two elements overlap at the node. In the case of a negative angle the seaming must be performed and nodes N_{i-1} and N_{i+1} are merged. The new node is denoted as node N_j and it is shown in Figure 5.12(b). By checking the angle and the number of attached elements according to the condition of Equation 5.14 further seaming is required at node N_j. The result of merging node N_{j-1} and N_{j+1} is shown in Figure 5.12(c). It can be seen that because of the seaming process the angle at the nodes (N_i, N_j and N_k) opens up which allows the formation of better quadrilateral elements.

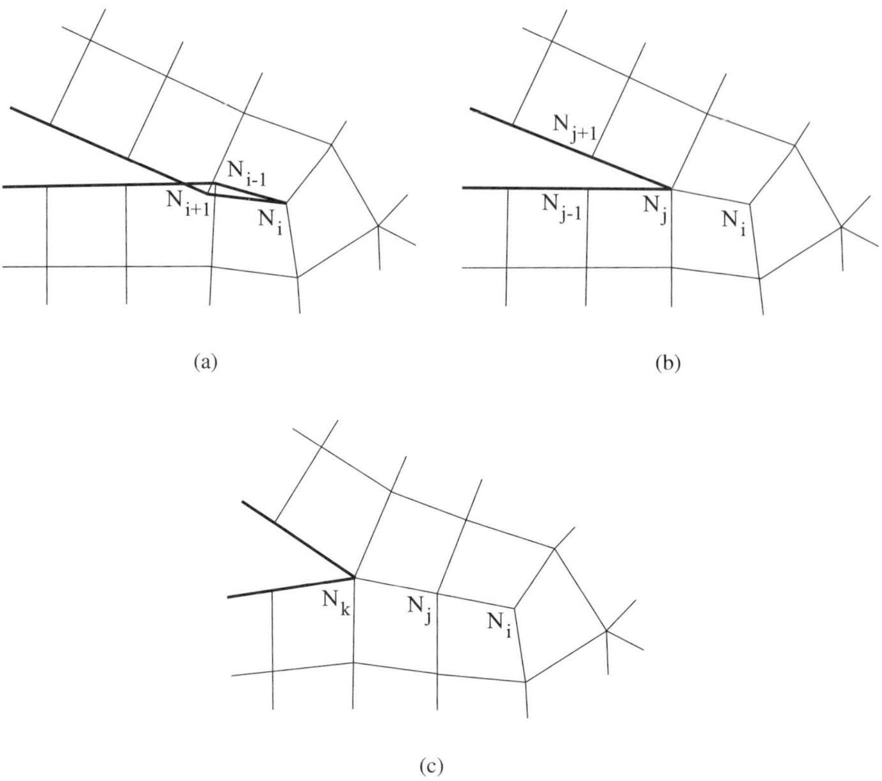

Figure 5.12: Steps of an interior node seaming, (a) the front overlaps with itself, (b) nodes N_{i-1} and N_{i+1} are merged into N_j, (c) the small angle at node N_j is eliminated by merging nodes N_{j-1} and N_{j+1}.

5.4.6.2 Transition seaming

The transition seaming process differs from the general case only in the lengths of the sides that are to be seamed. As can be seen in Figure 5.13(a), if node N_{i-1} and node N_{i+1} are to be merged, then the shaded element would be a poor quality quadrilateral element, since the side lengths of the element would be significantly different. To avoid this situation a wedge element is inserted as shown in Figure 5.13(b). The wedge element introduces two new nodes on the paving boundary. Nodes $N_i^{'}$ and $N_i^{''}$ are positioned at 1/3 and 2/3 length along the longer size. In the shaded element shown in Figure 5.13(b) node N_i is changed to node $N_i^{''}$. After the wedge element is inserted the seaming can be performed between node N_{i+1} and $N_i^{'}$. Figure 5.13(c) shows the state of the paving front after smoothing.

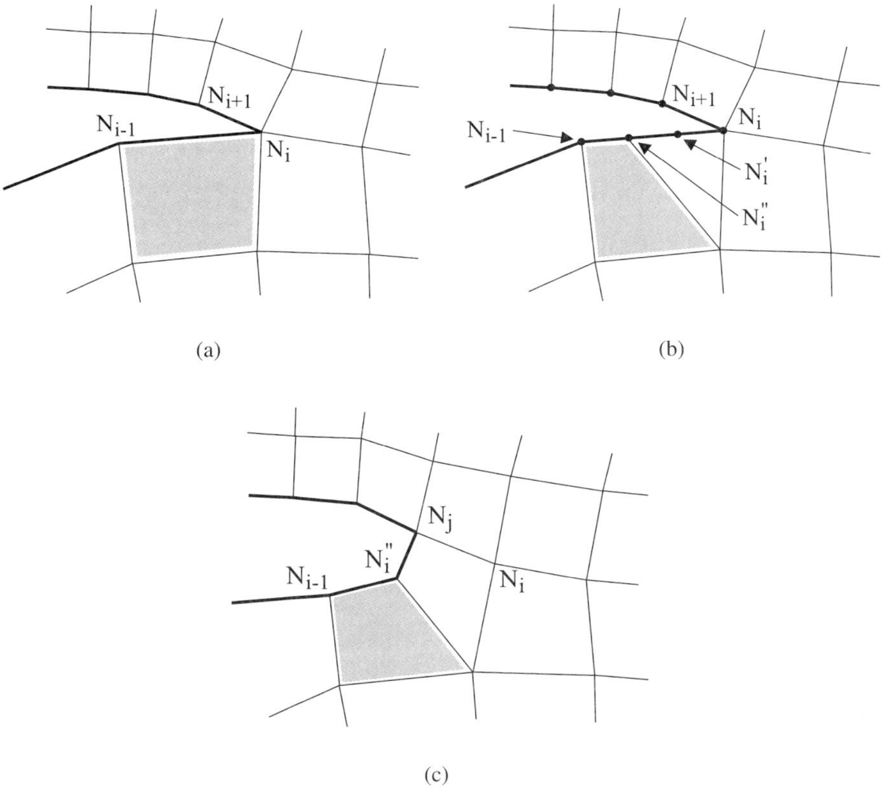

Figure 5.13: Transition seaming, (a) the length of the element sides at node N_i are different if it seamed it would form irregular quadrilateral element, (b) extra nodes N_i' and N_i'' are inserted to form a new quadrilateral element, (c) seaming is performed.

5.4.6.3 Seaming on mesh boundary

If a seam would be performed between node N_{i-1} and N_{i+1} in the situation shown in Figure 5.14(a) then a degenerate quadrilateral element would be formed (shaded in the figure). It is difficult to correct this element generation later, since mesh boundary nodes are not allowed to move. In this case, no seaming should be performed, but it should be delayed and the next portion of elements should be generated along the mesh boundary (see Figure 5.14(b). Thereafter the seaming can be performed as described in the previous Section 5.4.6.2. The final form is shown in Figure 5.14(c).

On the mesh boundary at sharp angles the formation of the elements is always similar to the case shown in Figure 5.14(c). However there are exceptions, because in the implementation the algorithm first checks if the seaming would produce a valid, good quality element. If the element is acceptable then the seaming solution is retained. If the element is unacceptable the seaming procedure is delayed.

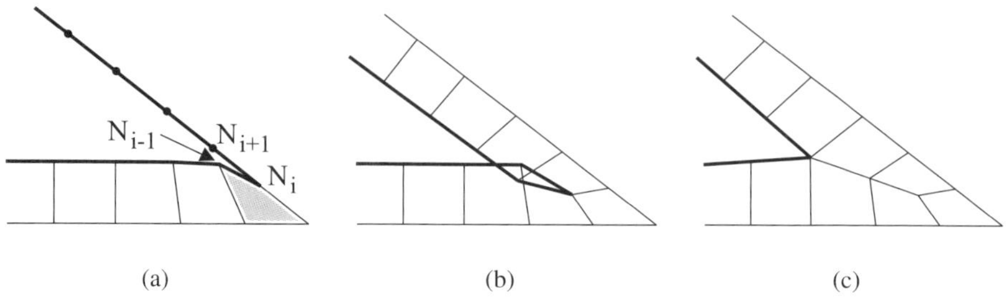

Figure 5.14: Seaming at a mesh boundary, (a) seaming is delayed to avoid the formation of a badly shaped quadrilateral element, (b) paving front overlaps itself, (c) seaming is performed.

5.4.7 Wedges and tucks

When the paving front contains a series of side nodes where the internal angle is non-convex or a series of side nodes where the internal angle is convex, the size of the elements can change dramatically from row to row. The size of the elements expands for the convex shape as shown in Figure 5.15(a) and it shrinks for the non-convex shape shown in Figure 5.16(a). To recognize this element formation and to help to correct the size of the elements, a technique, suggested by Blacker and Stephenson [6], can be used, where for the side expansion *wedges* are inserted and to avoid side shrinking *tucks* are formed.

5.4.7.1 Wedge insertion

The criteria for side expansion or for wedge insertion are that the paving front traverses more than 90 degrees and that in the particular section there are only side nodes where the internal angle is larger than 183 degrees.

The other criterion mentioned by Blacker and Stephenson [6] is the side expansion ratio which is defined as the ratio of the "required" length and the actual length of a segment on the paving front. The required length is calculated by Equation 5.13. This condition is not used in the current implementation. The reason is that, according to our experience, wedge insertion is used rarely especially with explicit size control and only in some specific cases. Thus, when wedge insertion should be used, for example in Figure 5.16, it is clear that expansion of the sides would happen. Though the expansion ratio may not reach the threshold level immediately, sooner or later some precautionary steps should be taken (such as inserting a wedge). The implementation presented here attempts to insert the wedge as early as possible therefore the first condition is deemed to be sufficient to detect the side expansion feature.

Theoretically a wedge should be inserted at every 90 degree section of the detected side expansion, but the number of available edges in the paving boundary also influences the number of inserted wedges. There should be at least two edges per wedge.

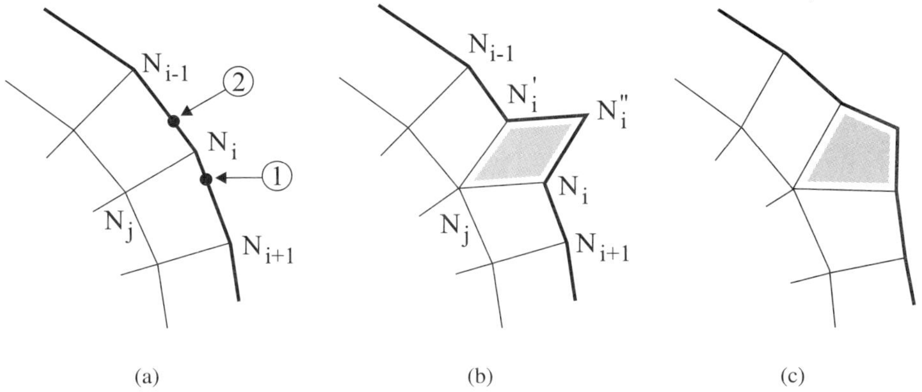

Figure 5.15: The wedge insertion technique, (a) node N_i is moved to position 1 and another node is created at position 2, (b) a new quadrilateral element is inserted, (c) afterwards the front is smoothed.

When the algorithm determines that a wedge should be inserted at node N_i (see Figure 5.15(a) then two new nodes are inserted into the paving front. First, node N_i is moved towards node N_{i+1} along their connecting line by 1/3 of the distance between the nodes into new position 1 shown in Figure 5.15(a). The first inserted node ($N_i^{'}$) will be placed at position 2 in Figure 5.15(a) which is 1/3 distance away from node N_i along the connecting line between node N_i and N_{i-1}. The position of the second inserted node $N_i^{''}$ is determind by the following formula:

$$V_i^{''} = V_i + V_i^{'} - V_j \tag{5.15}$$

where vector $V_i^{''}$ will point to the position of node $N_i^{''}$ and the other vectors point to the node denoted by the index of the vector. After the new points are inserted a new element is inserted as shown in Figure 5.15(b) by the shaded square. Figure 5.15(c) shows the final situation after smoothing, where the sizes of the edges on the paving front are reduced as a result of the effect of the two "pushed-in" edges.

5.4.7.2 Tuck formation

When a series of side nodes with angles less then 177 degrees is detected, then tucks should be formed to avoid further shrinking of element sizes. Blacker and Stephenson [6] suggested another condition but this is not used in the implementation here for the reasons discussed in Section 5.4.7.1. An example for the shrinking is shown in Figure 5.16(a). The angle traversed by the side nodes should be larger than 90 degrees and there should be four edges for each tuck. The tuck formation is most beneficial if it is inserted for every 90 degree section.

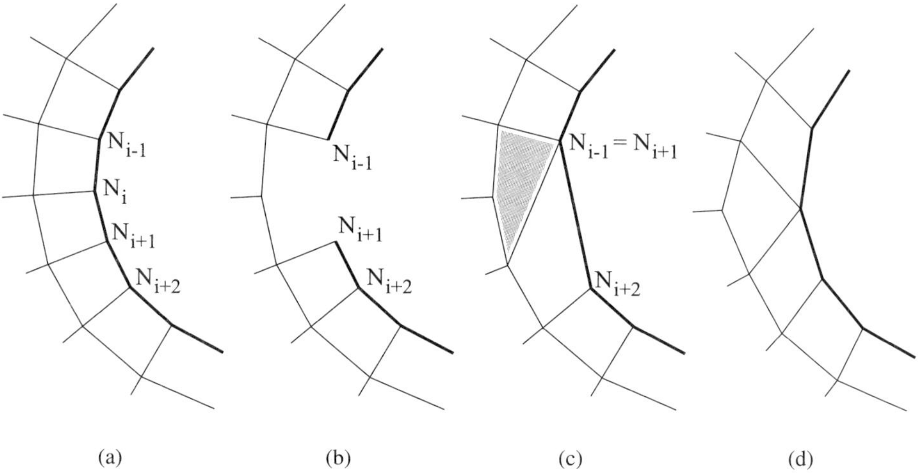

Figure 5.16: Tuck formation, a) contraction of size of element sides, (b) two quadrilateral elements are removed, (c) node N_{i+1} is merged with node N_{i-1} and a new quadrilateral element is formed, (d) the front is smoothed after tuck formation.

In the tuck formation node N_i and the connected two quadrilateral elements are deleted as shown in Figure 5.16(b). Node N_{i+1} is moved to the same location as node N_{i-1} and the shaded quadrilateral element is formed (see Figure 5.16(c). (Actually if only one element is deleted at the first step, then at this stage only the nodes of the other element should be changed in such a way that the shaded element is formed.) The final stage of the tuck formation with smoothing is shown in Figure 5.16(d) where the size of the edges on the paving boundary is increased as a result of the removed edges.

5.5 Row generation

This section covers the topic of the row generation procedure in detail, such as final node classification, recognition of primitive shapes and element generation strategies.

5.5.1 Final node classification

The first step is to classify all nodes according to Section 5.4.4. As described before, the classification can produce some ambiguous nodes which may be classified as being of one of two node types. The decision about the final type of these ambiguous nodes is based on whether they would produce regular or irregular nodes (see Figure 5.4) and in the case of an external boundary whether some primitive shapes could be recognised. In the case of the internal loops and when no primitive shape can be recognised for an external loop, the nodes are checked and if, by using one of their types, they would produce a regular node then that classification is accepted. If neither of the classifications would produce a regular node, the type is selected at random from the two possible types.

5.5.1.1 Recognition of primitive shapes

Three kinds of shapes can be recognised: rectangles, triangles and semi-circles (Figure 5.18). To recognize a primitive shape a global view of the current paving front should be obtained and an exhaustive search is performed to investigate all possibilities. A primitive shape is possible only when there are no row reversal nodes and row corner nodes on the paving front and the paving front is an external loop. In this case all nodes are classified temporarily as either a side node or an end node. An ambiguous corner node is always classified as a side node while an ambiguous side/end node can be considered as either a side node or as an end node. These ambiguous end nodes are called *candidate row end nodes*. To find a primitive shape a combinatorial search is performed using all end nodes and candidate end nodes.

For example, the paving front shown in Figure 5.17 has two end nodes, node 1 and 2, and four candidate end nodes (nodes 3, 4, 5, 6). To check whether the boundary is a rectangular primitive shape, shown by a rectangular shape with dashed line in Figure 5.17, nodes 1 and 2 are considered with another two candidate end nodes, such as nodes 3 and 4. In this case nodes 5 and 6 are considered as row side nodes. To find the best quality primitive shape all combinations of the candidate end nodes are tried with the row end nodes.

A shape will be accepted according to its quality and its validity. The validity of a shape is determined according to the rules given by Stephenson and Blacker [23]. Generally for all valid shapes it must be true that the number of edges must be even.

- A rectangle shape is valid if:

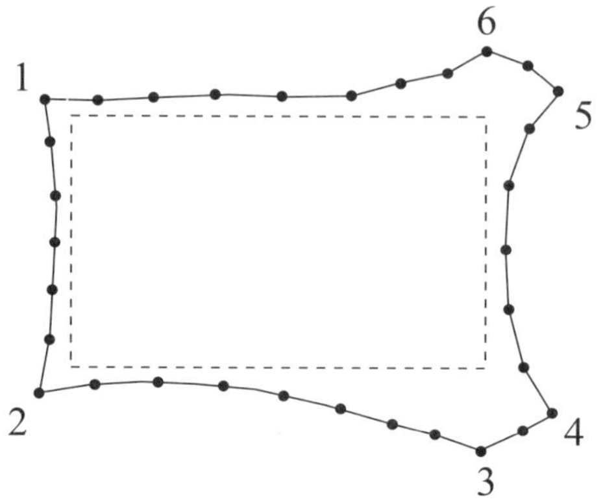

Figure 5.17: Example for primitive shape search.

- There are four row end nodes or in other words four sides in the shape.
- The sum of the number of edges on all sides is equal to or greater than four (see Figure 5.18)

$$n_1 + n_2 + n_3 + n_4 \geq 4 \tag{5.16}$$

- A triangle shape is valid if:
  - There are three end nodes which indcates that there are three sides in the shape.
  - Each side must have at least two edges

$$n_1 \geq 2 \tag{5.17}$$
$$n_2 \geq 2 \tag{5.18}$$
$$n_3 \geq 2 \tag{5.19}$$

  - The sum of the number of edges of any of the two sides is greater than or equal to the number of edges on the third side plus two (see Figure 5.18)

$$n_1 + n_2 \geq n_3 + 2 \tag{5.20}$$
$$n_2 + n_3 \geq n_1 + 2 \tag{5.21}$$
$$n_3 + n_1 \geq n_2 + 2 \tag{5.22}$$

- A semi-circle shape is valid if:
  - There are two row end nodes or in other words two sides in the shape.
  - The number of edges on the side with more edges must be greater than or equal to the number of edges on the other side plus four (see Figure 5.18)

$$n_1 \geq n_2 + 4 \tag{5.23}$$

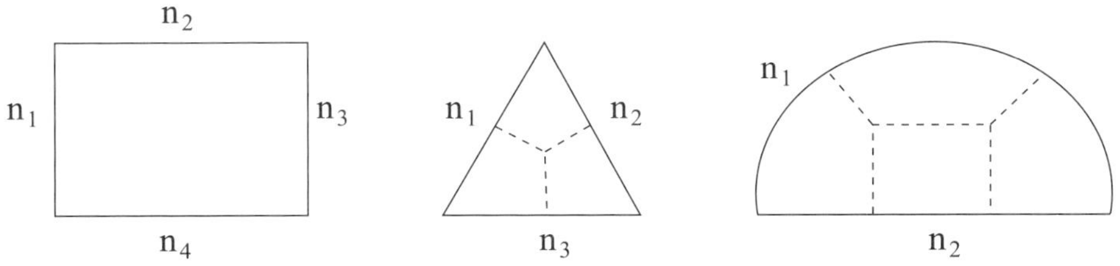

Figure 5.18: Primitive shapes. The figure also shows the number of generated irregular nodes.

A shape is accepted only when it is valid. The quality of a shape is determined by the number of irregular nodes that the paving process will produce using the current classification and the sum of the deviations of the angles from the "ideal" angle required by the classification of the nodes. For a row end node the "ideal" angle is 90 degrees and for a row side node it is 180 degrees. These choices for the angles of course favour the rectangle shape. On the other hand preferring the rectangular shape is beneficial, because if the classification of the nodes determined by a recognised primitive shape would be kept for all further element generation the number of irregular nodes would be zero for a rectangle, one for a triangle and two for a semi-circle shape as shown in Figure 5.18. A shape with fewer numbers of generated irregular nodes and smaller angle deviation is always accepted. An extra condition is that if the number of irregular nodes would be zero and the angular deviation is smaller than 45 degrees then the shape is also accepted.

When a valid and good quality shape is found, then all nodes on the paving boundary are finally classified according to the accepted shape. This shape classification is kept as long as the generation of the whole front has either finished or is interrupted. On the other hand when the whole front has been paved successfully using the classification of the recognised primitive, this shape classification is not inherited by the new front. In this way, for example, if a shape is recognised and paved as a triangle but later the shape becomes a rectangle then the method may correct itself.

5.5.2 Initial element generation

Generally when all the classification is finished and all nodes on the paving front are of one type the generation of the row of quadrilateral elements would start from an end node and finish at an end node. Blacker and Stephenson [6] identified three special cases when the row generation from an end node to an end node is not possible and they offered some solutions. Unfortunately they did not offer a solution for the situation shown in Figure 5.19 and it is not clear how the method would resolve this type of situation. To avoid the handling of all special cases in the implementation here the starting and closing criteria have been changed as described below.

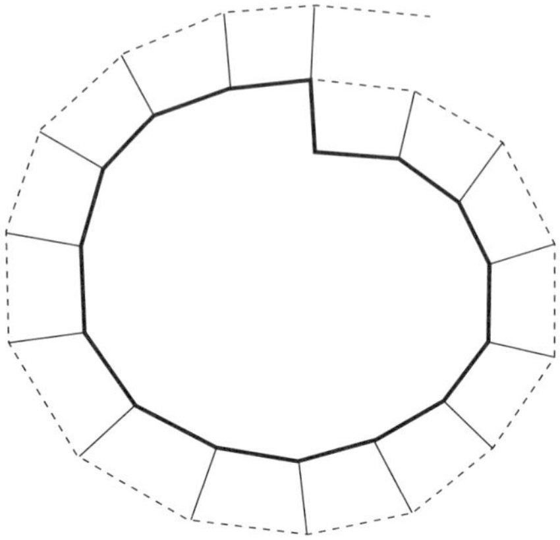

Figure 5.19: Special case.

The starting condition depends on the types of nodes which are available on the current paving front:

- If a primitive shape is recognised as described in Section 5.5.1.1 then the paving can start from any of the row end nodes.
- If there is at least one end node or candidate end node and there are some other type of nodes, then the row end node is selected as a starting node.
- If there is no row end node or candidate row end node, but there is at least one row corner or row reversal node then the starting point will be the node with the smallest internal angle.
- Similarly if there are only row side nodes on the paving front, then the node with the smallest internal angle is selected.

If the selected starting node is not an end node, then an extra node should be generated. For example when there are only row side nodes on the current paving boundary and the paving has to start at node N_i as it is shown in Figure 5.20(a) then two extra nodes are inserted on the current paving front. First node N_i is duplicated then node N_j is generated, but no elements (see Figure 5.20(b). In the figure the duplicate nodes N_i are pulled apart, but in reality they lie exactly at the same position. The newly created, duplicated node N_i is classified again by recalculating its internal angle and may serve as the closing row end point. Figure 5.20(c) shows the initial element as a shaded square with the direction of paving. A similar example is shown in Figure 5.21 for a row corner node.

In most cases this process ensures that there is a row end node to finish the row paving but to handle the situation in Figure 5.19 another condition is required. The row generation will be finished if the method encounters a newly created node. If this condition is not set, then the paving Figure 5.19 would continue as long as the spiral can be generated.

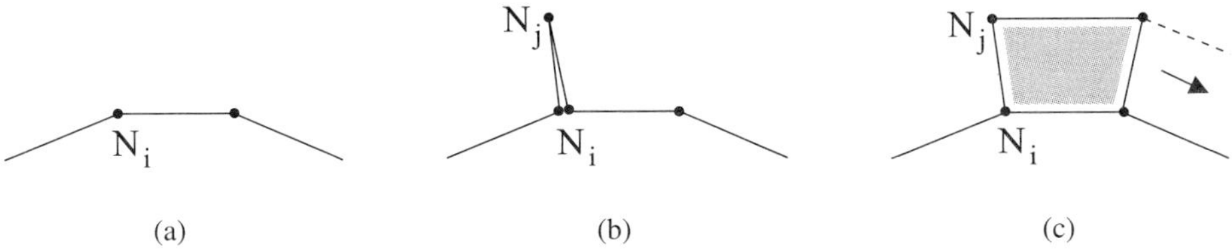

Figure 5.20: Initial element generation at row side.

5.5.3 Initial element generation: implementation

There are some special cases when the initial element cannot be generated because it would be an invalid quadrilateral element. An example is shown in Figure 5.22. The figure shows the paving direction and that invalid quadrilateral elements would be generated at the two row end nodes. The elements are invalid, as discussed in Section 5.4.1, as a result of their "flipped" shape.

In these cases the implemented procedure is the following. The algorithm first tries to start the paving, but it realizes that there is at present no valid start-up position. Then it marks the loop with the MUST_RESOLV bit and goes on to the next loop. When a marked loop is encountered then the quality requirements are relaxed during intersection resolution and "invalid" elements may also be accepted as a starting element.

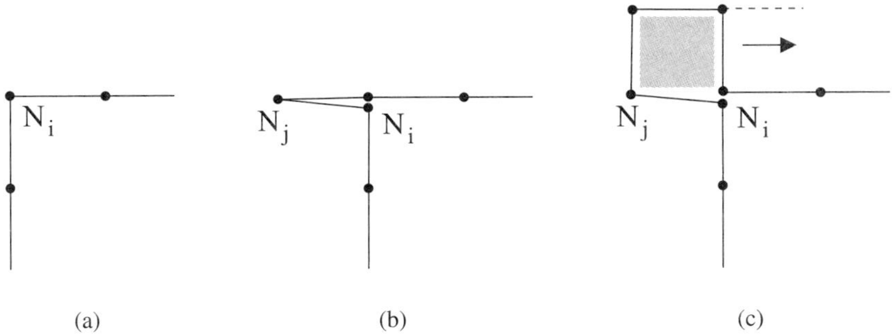

Figure 5.21: Initial element generation at a row corner node.

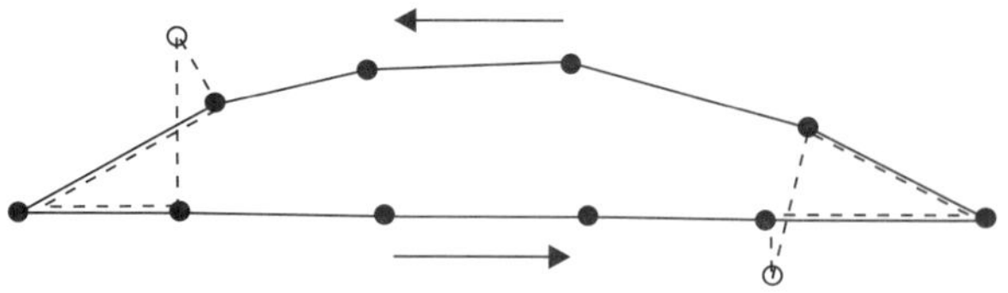

Figure 5.22: The paving front cannot be started as a result of invalid element generation.

The algorithm then relies on the smoothing process which will transform the invalid element into a valid one. If even this procedure fails, the algorithm goes through each node on the current paving front, tries to generate a quadrilateral element and tries to resolve the intersection or other problems. If it still cannot succeed the algorithm stops.

5.5.4 Element generation strategies

Blacker and Stephenson [6] described a row based paving approach. In this case a row of elements is inserted from a row end node to a row end node according to the rules of Section 5.4.5. Any intersection or other problems are resolved *after* the row has been generated. This means for example that if a row is placed over another one, the method does not detect this and the paving can "break-out". (A "break-out" happens when the method starts placing elements outside the area enclosed by the paving front.) On the other hand there are several possibilities to resolve an intersection by using not only the intersecting segments but segments adjacent to and further along the paving front.

White and Kinney [27] redesigned the row based paving approach for an element-by-element procedure. In the new method after an element is inserted it is immediately checked for intersection problems and quality. If the intersection cannot be resolved or the element is invalid or of poor quality the last step is undone and the paving continues at some other place. This approach helps to avoid "break-outs" and resolves intersection problems on the spot as they are created. Unfortunately in this case the number of ways to resolve an intersection is limited, since only the edges of the new element or the edges of the neighbouring element can be used.

In the current implementation the element-by-element approach, first proposed by White and Kinney [27], is adopted, with full undo capabilities. In this case there are two key functions:

- `FaceFactory`: This function (named after the description given by White and Kinney [27]) generates a new element at the current node on the active front. Each node remembers where the generation of elements has been stopped at the previous stage (see Section 5.10).
- `FactoryMonitor`: This function checks the quality of the generated ele-

ments immediately after creation.

In the FactoryMonitor function the first step is to check whether the generated quadrilateral element is valid. This step is skipped if the MUST_RESOLV bit is set for the loop (see also Section 5.5.3). Then the method tries to seam the newly created edges. If the seaming is successful the function returns telling the FaceFactory function that the element has been accepted and problems have been resolved. On the other hand if there was no seaming, then it checks the newly created segments with all paving front segments for proximity and intersection. If there are segments intersecting or within proximity then the algorithm tries to find the best match between the two segments. The best match occurs when the modified quadrilateral elements are still valid and their quality is above a certain threshold. When there is no intersection or segment in proximity the element is simply accepted. When two segments have been merged the function signals "element acceptance" and "problem resolution". The FaceFactory function relies on the output of the FactoryMonitor function. If the output of FactoryMonitor function is simply "element acceptance", then the generation can continue. If the output is "element acceptance" and "problem resolution" then the row generation stops and other adjustments are carried out on the paving front. The other possibility is that FactoryMonitor function signals that the element should be deleted, because it violates some criteria. In this case the FaceFactory function undoes the last node and element creation operations and the row generation is also stopped.

5.6 Intersection handling

The paving method implementation presented here assumes that there are no intersecting segments on the current paving front. To satisfy this condition every time an element is created the newly created segments are checked for intersection and proximity with all other paving front segments. The proximity is evaluated according to the following formula which is based on the criteria given by Cass *et al* [8]

$$P = \beta_1 \, P_{DIST} + \beta_2 \, P_{PERP} \tag{5.24}$$

where P_{DIST} is a value depending on the distance and P_{PERP} is a value depending on the angle between the two segments. P_{DIST} is calculated as

$$P_{DIST} = \frac{d}{S} \tag{5.25}$$

and

$$S = \gamma \, \frac{(l_1 + l_2)}{2.0} \tag{5.26}$$

where d is the shortest distance between the two segments, l_1 and l_2 are the length of the segments and γ is 0.45. The full derivation of γ can be found in reference [8] where S is called the *sticky space*. Basically the sticky space should express whether the two segments are close enough to merge them together or if there is enough space in between to insert another element. Unfortunately this condition assumes that the two segments are parallel and their end points are aligned. Since this is rarely the case, the P_{PERP} factor should also be calculated. The formula for P_{PERP} is

$$P_{PERP} = 1.0 - (e_1 \cdot e_2) \tag{5.27}$$

where e_1 and e_2 are the unit vectors of the two segments and the formula calculates the scalar product between the two segments. If the two segments are perpendicular then the value is 1.0 and if they are parallel then the value is zero.

The β_1 factor is 1.0 and β_2 is 0.25. The high value of β_1 factor expresses that regardless of the β_2 factor if the two segments are too far apart then they will not be further considered. If the value P is smaller than 1.0 the two segments are considered to be in close proximity.

In addition to the above proximity check a proper intersection check is also carried out between the new segments and all other segments. If an intersection is found, the method either resolves the problem by merging segments or the element has to be deleted.

After all intersecting and in proximity segments have been found all combinations of the segments are matched and the consequences of their merge is evaluated. It is verified that all affected quadrilateral elements are still valid and of good quality according to Section 5.4.2. Furthermore the merge would modify the paving fronts therefore the fronts are evaluated to ensure that the number of nodes on them remains even. If this latter condition is not satisfied, the merge is not possible.

After the appropriate segments have been found the algorithm carries out the merge and the control is passed to the adjustment functions as shown in Figure 5.1. If no merge is possible the element is deleted.

5.7 Smoothing

The smoothing operation is used regularly after almost every operation. It is applied not only after a row of elements has been generated, but also after seaming, wedge insertion, tuck formation or intersection resolution. This is a major difference compared to other mesh generator methods, where smoothing is usually applied only to the final mesh. As a result of this difference, the smoothing algorithms are specially designed for paving, since not only internal elements should be smoothed, but the "free" paving front as well. The smoothing procedures try to maintain the element size, ensure that the edges are perpendicular and generally make a smooth mesh. As a first step the nodes on the current paving front are smoothed, then these nodes are fixed and all internal nodes are smoothed.

5.7.1 Paving front smoothing

The nodes on the current paving front are smoothed by three different procedures:

- modified isoparametric smoothing,
- length adjustment and
- angular smoothing adjustment procedures.

The isoparametric smoothing (Δ_A) is applied to all nodes on the front, while the other two adjustments are applied only to nodes, where the number of attached elements (N_E) is two. In this case the average of the length (Δ_B) and the angular (Δ_C) adjustment is applied to the node. Summarizing

$$\Delta_i = \left\{ \begin{array}{ll} \Delta_A + \dfrac{(\Delta_B + \Delta_C)}{2} & \text{if } N_E = 2 \\ \Delta_A & \text{otherwise} \end{array} \right\} \tag{5.28}$$

5.7.1.1 Modified isoparametric smoothing

Let node N_i be on the current paving front, and attached to n quadrilateral elements then its new position, according to the modified isoparametric smoothing, is

$$V_i' = \frac{1}{n} \sum_{k=1}^{n} V_{ka} + V_{kc} - V_{kb} \tag{5.29}$$

where the vectors V_{ka}, V_{kb} and V_{kc} point to the nodes N_a, N_b and N_c of element k. Since the nodes of the elements are defined in an anti-clockwise manner node N_b is diagonally opposite to node N_i in element k. The vector to define the change of the position of node N_i is defined as

$$\Delta_A = V_i' - V_i \tag{5.30}$$

This type of smoothing tries to convert the quadrilateral elements into parallelograms but does not maintain the size of the elements.

However Kawamura *et al* [12] proposed a different formula which can be beneficial in situations where the size of the elements connected to node N_i is very different. They investigated crack propagation where very small elements around the crack are connected to larger surrounding elements. In this case Equation 5.29 distorts the small elements. The proposed formula is

$$V_i' = \frac{\sum_{k=1}^{n} \frac{1}{l_k} V_{ka} + V_{kc} - V_{kb}}{\sum_{k=1}^{n} \frac{1}{l_k}} \tag{5.31}$$

$$l_k = \frac{|V_{kc} - V_i| + |V_{ka} - V_i|}{2} \tag{5.32}$$

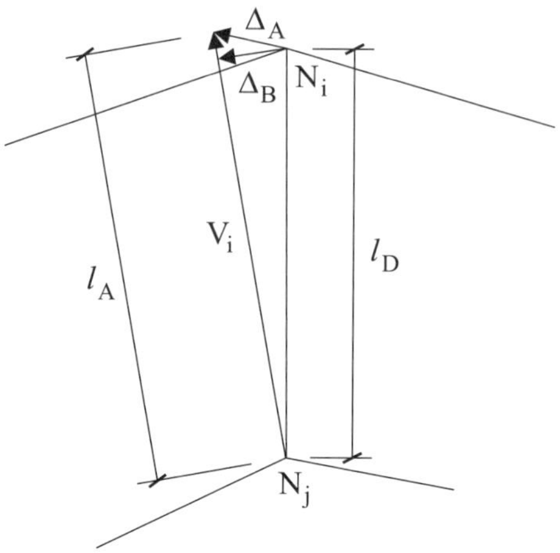

Figure 5.23: Length adjustment procedure.

where l_k is called the characteristic length. This formula has not programmed in the implementation used by the authors.

5.7.1.2 Length adjustment smoothing

Let node N_j be the base point for the projection to determine node N_i as shown in Figure 5.23 and P_i and P_j the vectors pointing to these points. Further more let l_D be the length which was used to generate node N_i according to Equations (5.4), (5.5) and 5.8 or alternatively Equation (5.13) and let l_A be the length between the base point N_j and node N_i after applying the isoparametric smoothing, Δ_A. Then the change of the position of node N_i for the length adjustment smoothing is defined as

$$\Delta_B = P_j - P_i + (\Delta_A + P_i - P_j)\frac{l_D}{l_A} \tag{5.33}$$

This procedure tries to adjust the length of vector V_i and maintain the required element size.

5.7.1.3 Angular adjustment smoothing

Let node N_j be the base point for the projection to obtain node N_i as shown in Figure 5.24 and node N_{i-1} and N_{i+1} be adjacent to N_i. The vectors pointing from node N_j to node N_{i-1}, N_i and N_{i+1} are V_{i-1}, V_i and V_{i+1} respectively. Now let vector V_{b1} bisect the angle between vector V_{i-1} and V_{i+1} and then vector V_{b2} bisect the angle between vector V_{b1} and V_i. The direction of vector V_{b2} determines the new direction of the vector V_i. The intersection point of vector V_{b2} and the line between node N_{i-1}

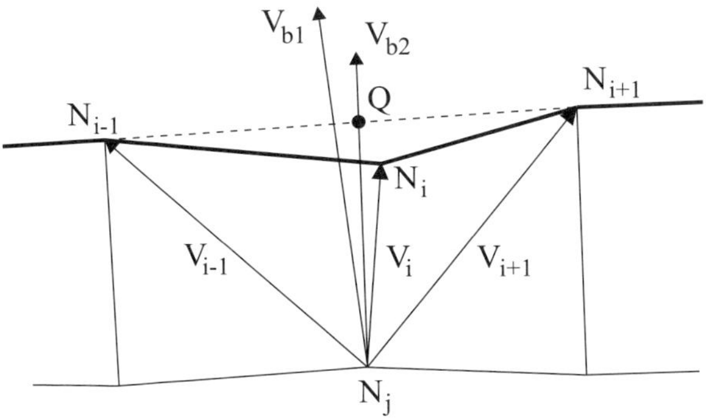

Figure 5.24: Angular adjustment procedure.

and N_{i+1} is defined as Q which is also shown in Figure 5.24. The length of the vector from node N_j to Q is defined as l_Q and the original projection length which was used to generate node N_i is l_D. The length of the vector V_{b2} is

$$|V_{b2}| = \left\{ \begin{array}{c} \dfrac{(l_Q + l_D)}{2} \text{ if } l_D > l_Q \\ l_D \text{ otherwise} \end{array} \right\} \tag{5.34}$$

Therefore the new position of node N_i due to angular smoothing is

$$\Delta_C = V_{b2} - V_i \tag{5.35}$$

This procedure tries to make element edges perpendicular to each other and also helps to adjust the lengths of the vector V_i to provide a smooth paving front.

5.7.2 Interior node smoothing

Let node N_i be the node to smooth and let V_j be the vector pointing from node N_i to an attached node N_j as shown in Figure 5.25. Blacker and Stephenson [6] suggest that this vector should be modified by the angular smoothing vector Δ_C for nodes on the mesh boundary to ensure perpendicularity. In this implementation this modification is not applied to vector V_j. If there are n attached elements to node N_i then the change of the position of node N_i is defined as

$$\Delta_i = \frac{\sum_{j=1}^{n} |V_j| \, V_j}{\sum_{j=1}^{n} |V_j|} \tag{5.36}$$

The procedure attempts to move node N_i to an equal distance from all other nodes.

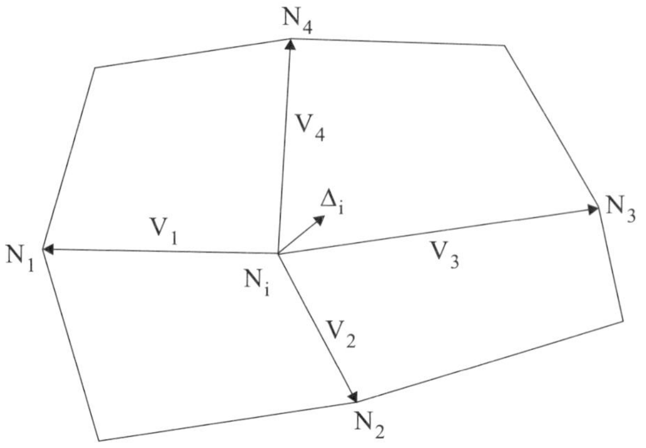

Figure 5.25: Interior node smoothing.

5.7.3 Further comments about smoothing

The smoothing operations can be computationally very expensive, but they are essential for the method to succeed, therefore some modification of the algorithm is required to speed it up. Some suggestions relating to speeding up of the process are discussed by Blacker and Stephenson [6].

When the paving front advances into the region to be paved the nodes which are left behind (internal nodes) do not change greatly after a certain distance from the current paving front. Therefore, it is beneficial to the speed and does not significantly reduce the quality of the mesh if the smoothing is only applied to nodes which are close to the current paving boundary. In the implementation presented here this means that only nodes which are three element levels deep from the current paving front are smoothed, and not the whole mesh.

The nodes on the current front and the orientation of the corresponding elements are known. It is therefore possible to determine the node below the node on the front, then the node below the previous node. This is not perfect, as one node can occur more than once but in this way we go "down" identifying nodes to three levels deep.

5.8 Loop closure

When the number of nodes is less than six in the current paving front then a closing procedure is applied to the front according to the number of nodes. If the number of nodes in the loop is only two, the loop is simply deleted. In the case of four remaining nodes an extra quadrilateral element is created from the nodes, then the loop is deleted. In the case of six nodes in the loop, extra nodes and elements are inserted into the mesh according to the configuration of the loop. To determine the configuration of the loop all row end nodes are identified and the number of edges between the end nodes are also counted. The classification procedure is exactly the same as for the normal paving procedure. Depending on the number of end nodes and the number of edges between them (or alternatively considering the loops as primitive shapes) several loop closing schemes have been described by Blacker and Stephenson [6] and they are summarised in Figure 5.26. Some of the quadrilateral elements in the table are of bad quality, but since these closures occur mostly inside the mesh the smoothing process will generally adjust the quality of the elements.

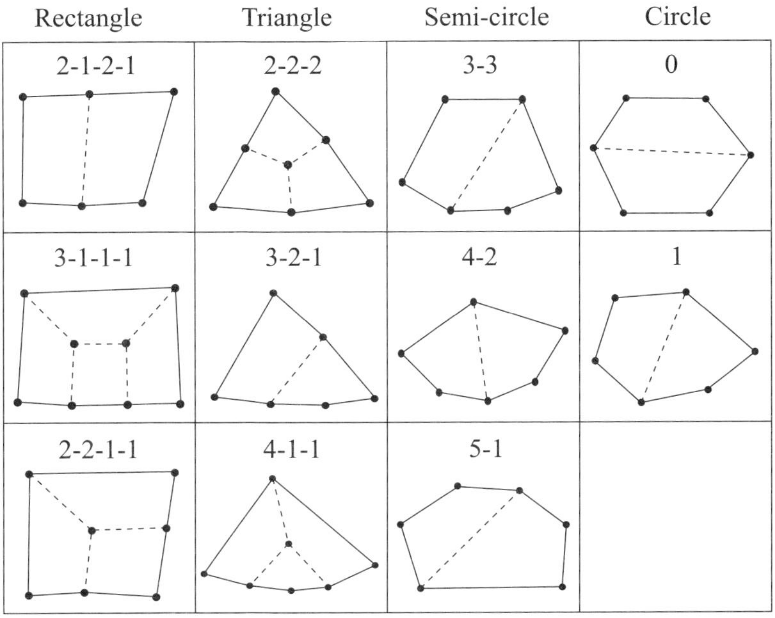

Figure 5.26: Paving front closing strategies depending on configuration of the number of edges and end nodes.

5.9 Final mesh modifications

After the full domain is meshed some badly shaped elements can still exist. To remove these elements several methods can be used which are described in Chapter 9. Methods to merge quadrilateral elements or modify their topology to improve the mesh quality were proposed by Blacker and Stephenson [6], Cheng [9] and Kawamura [12].

Kinney [14] proposed some other clean-up procedures which are implemented in CUBIT [3,21] but not in the implementaion presented here. These procedures consider certain patterns of quadrilateral elements and alter their connectivity to achieve better quality. Other smoothing techniques for quadrilateral meshes have been developed by Staten and Canann [22].

5.10 Data structures

The representation of the whole mesh in the program is encapsulated in one structure. It contains the node, element, boundary and front descriptions. The coordinates of the nodes and the nodes which define the quadrilateral elements are stored in two dimensional matrices. The size of the matrices is increased as new nodes and elements are added. This incremental memory allocation is not carried out in a one-by-one fashion. For example, when a new element should be added to the element definition matrix and there is no more space in the matrix then the program allocates a predefined amount of memory which can hold more than one element definition. In this way no allocation is necessary until the newly allocated space is full.

An extra matrix provides some information about the relation between the nodes and elements. In this matrix each row corresponds to a node in the mesh and within the row the indices of all quadrilateral elements which contain the particular node are listed. Another project [13] which implemented the paving procedure used a quad-edge data structure [10].

In the implementation presented in this book the paving fronts are represented as circular double linked lists. The list data structure stores a reference to the current node, the number of nodes in the list, the type of the front (external or internal) and whether the shape of the boundary is primitive (see in Section 5.5.1.1). Each element of the double linked list is a node on the paving front. A node stores a reference to its neighbours on the front, some flags indicating the node type, the index number of the node referring to the node definition matrix, the internal angle, the required length at the node and some "state" variables. The "state" variables are used when at the current node several quadrilateral elements should be generated. Since this implementation follows the element-by-element approach (see Section 5.5.4) several operations must be carried out after the creation of an element, therefore the current node should remember the current state of the generation to be able to resume its process at a later stage.

5.11 Concluding remarks

In this chapter, the paving method has been described and the implementation discussed. Examples of this method will be given in Chapter 10.

Bibliography

[1] Aleksa, M., Russenschuck, S., Vollinger, C., "Parametric quadrilateral meshes for the design and optimization of superconducting magnets", Technical Report CERN-LHC-Project-Report-448, CERN, Geneva, Switzerland, 2000, also presented at the 6th International Computational Accelerator Physics Conference, Darmstadt, Germany.

[2] Blacker, T.D., "Automated quadrilateral surface discretization method and apparatus usable to generate mesh in a finite element analysis system", United States Patent 5,315,537, May 1994.

[3] Blacker, T.D., *CUBIT Mesh Generation Environment, Volume 1: User's Manual*, Sandia National Laboratories, Albuquerque, New Mexico, 1994, SAND94-1100.

[4] Blacker, T.D., "FASTQ Users Manual Version 1.2", Technical Report SAND88-1326, Sandia National Laboratories, Albuquerque, New Mexico, 1994.

[5] Blacker, T.D., Stephenson, M.B., "Paving: A new approach to automated quadrilateral mesh generation", Technical report, Sandia National Laboratories, Albuquerque, New Mexico, 1990, SAND90-0249.

[6] Blacker, T.D., Stephenson, M.B., "Paving: A new approach to automated quadrilateral mesh generation", *International Journal for Numerical Methods in Engineering*, 32: 811–847, 1991.

[7] Blacker, T.D., Stephenson, M.B., Canann, S., "Analysis automation with paving: A new quadrilateral meshing technique", *Advances in Engineering Software*, 56: 332–337, 1991.

[8] Cass, R.J., Benzley, S.E., Meyers, R.J., Blacker, T.D., "Generalized 3-D paving: An automated quadrilateral surface mesh generation algorithm", *International Journal for Numerical Methods in Engineering*, 39: 1475–1489, 1996.

[9] Cheng, B., *Parallel adaptive finite element mesh generation*, M.Phil. Thesis, Heriot-Watt University, Edinburgh, UK, 2001.

[10] Guibas, L., Stolfi, J., "Primitives for the manipulation of general subdivisions and the computation of Voronoi diagrams", *ACM Transactions on Graphics*, 4 (2): 74–123, 1985.

[11] Jang, C.D., Moon, S.C., Ko, D.E., "Acquisition of line heating information for automatic plate forming", in *Proceedings of Ship Structure Symposium 2000*, pages 203–208, 13th-14th June 2000.

[12] Kawamura, Y., Mu, Y., Sumi, Y., "Development of an automatic quadrilateral mesh generation for the simulation of curved crack growth", *Transactions of Japan Society for Computational Engineering and Science*, 1999, Paper no. 19990024.

[13] Kim, J.H., "Quadrilateral mesh generation using paving method", http://madmax.me.berkley.edu/j̃aeho/classes/cs294-5/proj2.html, August 1999, CS294-5 Project 2.

[14] Kinney, P., "CleanUp: Improving Quadrilateral Finite Element Meshes", in *Proceedings of the 6th International Meshing Roundtable*, pages 449–461, 1997.

[15] Ladeinde, F., Alabi, K., "A high-quality advancing front technique using recursive domain sub-division", in *Proc. 6th Int. Conf. Numerical Grid Generation in Computational Field Simulations*, pages 431–446, July 1998.

[16] Lober, R.R., Tautges, T.J., Vaughan, C.T., "Parallel paving: An algorithm for generating distributed, adaptive, all-quadrilateral meshes on parallel computers", Technical Report SAND97-0545, Sandia National Laboratories, Albuquerque, New Mexico, 1997.

[17] Nowottny, D., "Quadrilateral mesh generation via geometrically optimized domain decomposition", in *Proceedings of the 6th International Meshing Roundtable*, pages 309–320, 1997.

[18] Owen, S.J., Jones, N.L., Holland, J.P., "A comprehensive modeling environment for the simulation of groundwater flow and transport", *Engineering with Computers*, 12, 1996.

[19] Owen, S.J., Staten, M.L., Canann, S.A., Saigal, S., "Q-Morph: An indirect approach to advancing front quad meshing", *International Journal for Numerical Methods in Engineering*, 44: 1317–1340, 1999.

[20] Rees, M., "Combining quadrilateral and triangular meshing using the advancing front approach", in *Proceedings of the 6th International Meshing Roundtable*, pages 337–348, 1997.

[21] Sjaardeam, G.D., *CUBIT Mesh Generation Environment, Volume 2: Developers Manual*, Sandia National Laboratories, Albuquerque, New Mexico, 1994, SAND94-1101.

[22] Staten, M.L., Canann, S.C., "Post refinement element shape improvement for quadrilateral meshes", in *AMD-Vol. 220, Trends in Unstructured Mesh Generation*, pages 6–16. ASME, July 1997.

[23] Stephenson, M.B., Blacker, T.D., "Using conjoint meshing primitives to generate quadrilateral and hexahedral elements in irregular regions", in *ASME, Computers in Engineering Conference*, pages 163–172, 1989.

[24] Stephenson, M.B., Canann, S.A., Blacker, T.D., "Plastering: a new approach to automated 3D hexahedral mesh generation", Technical Report SAND89-2192, Sandia National Laboratories, Albuquerque, New Mexico, 1992.

[25] van Rens, B.J.E., *Finite element simulation of the aluminum extrusion process: Shape prediction for complex profiles*, Chapter "Meshing the surfaces - Paving generator", pages 48–53, Universiteitsdrukkerij TU Eindhoven, Eindhoven, 1999.

[26] van Rens, B.J.E., Brokken, D., Brekelmans, W.A.M., Baaijens, F.P.T., "A two dimensional paving mesh generator for triangles with controllable aspect ratio and quadrilaterals with high quality", *Engineering with Computers*, 14(3): 248–259, 1998.

[27] White, D.R., Kinney, P., "Redesign of the paving algorithm: Robustness enhancements through element by element meshing", in *Proceedings of the 6th International Meshing Roundtable*, pages 323–335, 1997.

Chapter 6

Two-dimensional Delaunay triangulation

6.1 Introduction

In this chapter the Delaunay triangulation method is used for the generation of two-dimensional triangular element meshes. This approach to mesh generation results in unstructured meshes and can also be used for refinement of existing meshes.

6.1.1 The history of Delaunay triangulation

Constructing triangulations has interested mathematicians and applied mathematicians throughout history. Boris Delaunay's suggested triangulation however only appeared in 1934 [9]. It was based on the early work of Dirichlet [10] and Voronoï [28] who showed how to construct proximity diagrams for a series of arbitrary vertices in two or three dimensions. In two dimensions the Voronoï diagram of a group of vertices divides two-dimensional space into regions bound by straight line segments. Each region, or Voronoï cell belongs to one vertex. The cell contains that part of the two-dimensionsional plane which is closer to the vertex that it is associated with, than to any other vertex in the group. Figure 6.1 shows the Voronoï diagram of a group of eighteen vertices with dotted lines. Seed [25] has shown that the conversion between the Voronoï diagram and the Delaunay triangulation (and *vice versa*) can be undertaken in linear time.

Delaunay showed that the Voronoï diagram could be converted into a unique triangulation of the vertices. This triangulation is referred to as the dual of the Voronoï diagram. To construct the dual of the Voronoï diagram it is sufficient to connect each two vertices, whose Voronoï cells are neighbours, by a new edge. The result of the conversion is the Delaunay triangulation (DT). Figure 6.1 shows the DT of the same group of 18 vertices with solid lines.

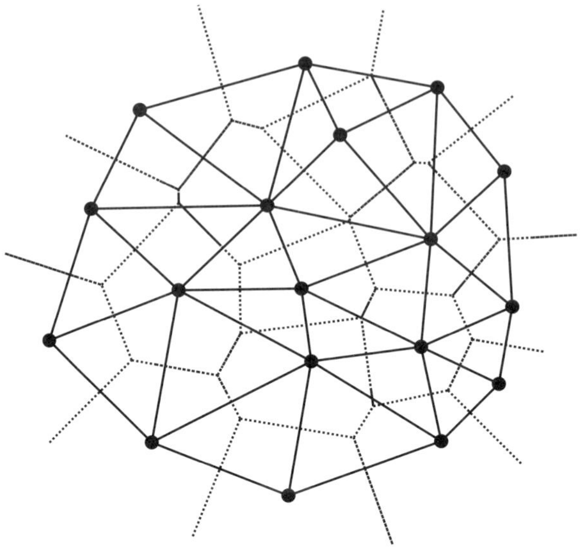

Figure 6.1: The Voronoï diagram (in dotted lines) and the Delaunay triangulation (in solid lines) of a group of eighteen vertices.

6.1.2 Delaunay based algorithm development

Several algorithms were developed for the construction of the DT of a group of vertices. A short overview is given here whereas more details on the different categories of algorithms will follow in Section 6.1.5. Early algorithms for the construction of the DT started from the Voronoï diagram, of which the construction was well known. Efficient ways of converting the diagram into a triangulation were developed.

In the seventies however Lawson [16] introduced a new technique called *topological flipping* which could be used to obtain the DT directly from any arbitrary triangulation. There was no longer a need to construct the Voronoï diagram first. As an alternative to Lawson's technique Green and Sibson [14] devised an incremental point insertion algorithm which enables the construction of the DT of a group of vertices by adding one extra vertex at the time to a gradually growing triangulation starting from a single triangle. The key to the scheme is that at every step the growing triangulation remains the *unique* DT of all the vertices that have been used. Bowyer [4] and Watson [29] independently extended this incremental algorithm to three and more dimensions.

Non-incremental methods were also developed, such as the *divide-and-conquer* method [18,26] and *sweepline* methods [12,23] which provided valuable alternatives.

6.1.3 The success of Delaunay triangulation for unstructured mesh generation

The use of Delaunay triangulation for mesh generation only appeared in the early eighties. DT was gradually extended from a mathematical method for triangulation to a robust and highly efficient technique for the generation of computational meshes. The triangulation techniques however had to be combined with a fast and robust scheme for the creation of mesh points. *Dependant* and *independant* point creation schemes were developed. In the dependant schemes the location of the inserted point is determined on the basis of connectivity of the triangulation. Independant schemes use the Delaunay triangulation technique as a triangulation method only. Points in these techniques are created in a separate procedure which is not related to the triangulation.

6.1.4 Fundamental properties of the Delaunay triangulation

If the Delaunay triangulation of a group of vertices is defined as the dual of the Voronoï diagram of that group, then the fundamental property of the DT is its uniqueness. This is because the Voronoï diagram of a group of vertices is unique and because the construction of the dual of a graph is a unique and reversible procedure.

Delaunay also demonstrated another very important property of the DT. He proved that the Delaunay triangulation satisfies the *empty circle criterion*. This means that the circumcircle around every triangle of the DT contains no vertices of the triangulation other than the three vertices that define the triangle. This remarkable property was proved in a different way by Lawson [16].

Another interesting property which explains partially the success of the method for mesh generation is the *min-max* criterion. First noted by Lawson [16] this criterion explains that of all possible triangulations of a group of vertices, the DT is the one that maximizes the minimum angle in the triangulation. Related to this property it can also be proved that the DT minimizes the largest circumcircle that can be constructed around any triangle.

6.1.5 Algorithms

As already described in Section 6.1.1 a classification can be made between incremental and non-incremental methods to construct the Delaunay triangulation of a group of vertices. Non-incremental methods require all vertex positions to be known in advance. The incremental methods are often referred to as point insertion schemes as vertices are added to the triangulation one at the time.

6.1.5.1 Non-incremental Delaunay triangulation methods

The divide-and-conquer strategy is a commonly used strategy to solve computational tasks such as triangulation. The Delaunay triangulation of a set of vertices is recursively constructed from two triangulations of subsets of the vertices.

The crux however is the merge step. Shamos and Hoey [26] suggested the first divide-and-conquer algorithm for constructing the Voronoï diagram in $O(N \log N)$ complexity. Green and Sibson [14] came up with an alternative approach which is of $O(N^2)$ complexity in the worst case. The Delaunay triangulation can be reconstructed from the Voronoï diagram in $O(N)$ complexity. Lee and Schachter [18] finally compared the algorithms and offered their own schematic implementation.

The $O(N \log N)$ sweepline algorithm introduced by Fortune [12] starts with sorting all the vertices that are to be triangulated. The algorithm then sweeps a line up in the y direction through the vertices. Whenever the line passes a new vertex it is connected to the triangulation which exists only below the sweepline. The choice of appropriate data structures is crucial to the efficiency of this algorithm. An alternative implementation was described by O'Rourke [23].

If the non-incremental methods are adopted for the purpose of mesh generation an external node creation scheme is required to provide the locations of the mesh nodes. This is often referred to as *automatic point creation* and *local reconnection* [19, 20]. The cited references refer to this as automatic point *insertion* instead of automatic point *creation*. Throughout this book however *creation* will be defined as the determination of a mesh node's coordinates whereas *insertion* will be used to describe the changes to mesh connectivity that are required to make the node part of the mesh.

The mesh nodes can also be created using advancing-front-type point placement as demonstrated by Mavriplis *et al* [21]. This method combines the triangle quality advantages of advancing front methods with the robustness and solid mathematical basis of Delaunay triangulation.

6.1.5.2 Incremental Delaunay triangulation methods

In the incremental methods each vertex is inserted into the triangulation individually. The first problem that arises is how to find which triangle of the current triangulation this new vertex belongs to. This triangle, once it is found, will be called the initial triangle. Rather than exhaustively checking all triangles in the mesh a searching algorithm referred to as the *walking algorithm* was developed by Borouchaki and Georges [3, 13]. This algorithm is explained in detail in Section 6.6.1.

The next challenge is to reconnect the new vertex to the triangulation while maintaining the Delaunay properties of the mesh. Two solutions have emerged in the literature:

- Lawson's edge-flipping algorithm [16] which unfortunately does not extend very well into three dimensions.

- The insertion polygon method described by Green and Sibson [14] which was generalized for three dimensions by Bowyer [4] and Watson [29].

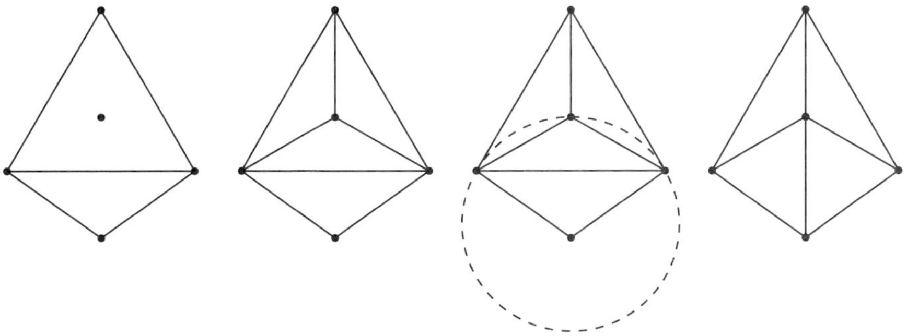

(a) (b) (c) (d)

Figure 6.2: The edge-flipping algorithm.

The edge-flipping algorithm is demonstrated in Figure 6.2. In Figure 6.2(a) a new vertex is shown which has to be inserted into an existing mesh consisting of two triangles. The algorithm starts with finding the initial triangle and splitting it into three new triangles, as shown in Figure 6.2(b). The circumcircle of any of these three new triangles however might contain the opposite corner node of a neighbouring triangle as shown in Figure 6.2(c). If this is the case, a flip of diagonals is carried out which is shown in Figure 6.2(d). For every new triangle created by flipping a circumcircle test also has to be carried out. As long as local violations of the Delaunay empty circle property are detected the edge flipping continues. Lawson proved that the flipping will finally stop and that the resulting triangulation will be the Delaunay triangulation.

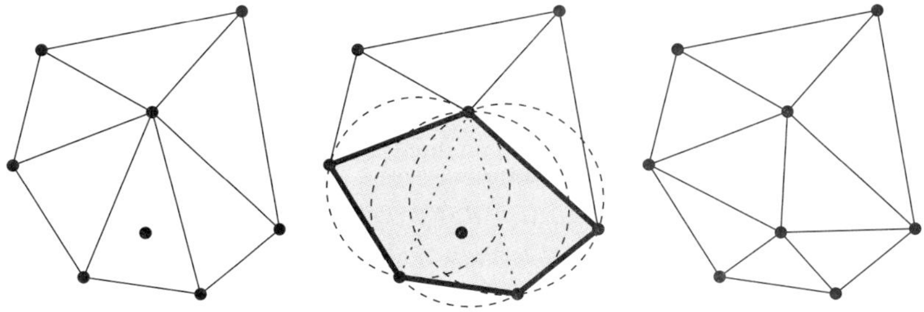

(a) (b) (c)

Figure 6.3: The insertion polygon method.

The insertion polygon method is demonstrated in Figure 6.3. The method first identifies all triangles (in the vicinity of the initial triangle) that have the property that the new vertex falls inside their circumcircle. In Figure 6.3(b) all these triangles are shaded gray. The enclosing polygon of these triangles is referred to as the insertion polygon, shown in Figure 6.3(b) with a thick line. In a second step all these triangles are removed from the triangulation. The new vertex is then reconnected to all vertices of the insertion polygon, as shown in Figure 6.3(c). The determination of the insertion polygon is unfortunately numerically sensitive. To make it more robust either more checking and correction procedures [3] have to be applied or adaptive precision floating-point arithmetic and fast robust geometric predicates [27] are required.

In this book the edge-flipping algorithm will be used for two dimensional Delaunay triangulation. The three dimensional equivalent of the insertion polygon method, the insertion polyhedron method will be applied for Delaunay tetrahedralization which will be described in Chapter 7.

6.1.6 Automatic point creation schemes

Incremental methods are extremely convenient for automatic mesh generation. The location of the mesh points will be determined from the connectivity of the mesh at different stages in the triangulation procedure. In general the mesh generation procedure includes the following stages:

- Perform an initial triangulation of the nodes that define the coarse mesh.
- Refine the boundaries of the domain until the nodes along the boundary obtain the required level of spacing. Insert all the boundary refinement nodes into the triangulation.
- Keep inserting field refinement nodes inside the triangles of the boundary node triangulation until all triangles satisfy the mesh criteria.

The boundary refinement step is fairly straightforward. Boundary refinement nodes are created along each segment of the boundary definition. It is important to note that when meshes for non-convex domains are considered these segments may not be part of the triangulation of the initial coarse mesh nodes. The Delaunay triangulation of a series of nodes does not give any guarantee as to which node connecting segments are part of the triangulation. The boundary segments may or may not be included in the triangulation. Boundary refinement nodes may therefore be inserted along a segment that only exists in the boundary definition but not in the triangulation.

The placement of field refinement nodes is however a unique step in the automatic point creation procedures. The following methods were suggested in literature:

- Chew invented two refinement algorithms. The first [6] is suitable for producing meshes with a uniform mesh size. His second algorithm [7] is for meshes with variations in the required mesh sizes over the mesh domain. His algorithm inserts new refinement nodes at the centers of the circumcircles of certain poor quality triangles. The quality of the triangles is judged by the ratio of their circumradius to shortest edge length. It has been proved that a uniform mesh generated with this algorithm contains no triangles with angles less than 30°.

- Ruperts algorithm [24] for field refinement nodes was specially developed for the generation of non-uniform meshes. In his implementation new nodes are also inserted at the circumcenters of poor quality triangles. The definition of the term *poor* however is more complex. According to Shewchuk [27] it is probably the first theoretically guaranteed meshing algorithm to be truly satisfactory. It also produces typically fewer triangles than Chew's second algorithm.

- Borouchaki and George [2, 13] suggested a method whereby the internal edges generated by the triangulation of the boundary refinement nodes are refined in the same way as if they were boundary segments. The new field refinement nodes however are filtered by a grid before final acceptance and insertion into the triangulation.

- Weatherill demonstrated a method [31, 32] which inserts field refinement nodes at the centroids of the unsatisfactory triangles. The new nodes are required to pass certain criteria before being accepted. This method will be explained in greater detail in Section 6.7.

6.1.7 Boundary recovery

Probably the greatest disadvantage of Delaunay triangulation is that for non-convex domains the resulting triangulation may not conform to the boundary. This means that boundary segments might not be present in the triangulation of all boundary nodes. The recovery of these boundary segments has challenged many researchers. Solutions to this difficulty can be divided into two categories:

- Those solutions that maintain the Delaunay properties of the mesh by making alterations that do not sacrifice the empty-circle criterion.

- And the solutions that break the Delaunay integrity of the triangulation under the condition that no further nodes need to be inserted.

Both these solutions will be discussed in Section 6.8.1.

Alternatively a new kind of triangulation has been developed which integrates boundary conformity as an essential feature. It is called the *Constrained Delaunay Triangulation* (CDT). First developed by Lee and Lin [17], in 1986, it is still limited to two dimensionsional mesh generation. Weatherill gave a recent overview of alternative approaches for boundary recovery in both two and three dimensions [30].

6.2 Data structures

Most programming attempts of geometry related programs fail or become incredibly complicated when data structures are not carefully designed. Mesh generation applications usually have considerable memory requirements as the number of elements in the meshes to be generated start to increase.

Search procedures can often be avoided by constructing lookup tables which store replicated data in different orders. A trade-off must be made between the resources taken by the extra memory required and the time lost by extra search cycles.

The most important data structure in any mesh generation program is obviously the mesh structure. How a mesh is defined and stored will undoubtedly influence the mesh generation code. In the authors' code the mesh definition follows the structure defined in the development library presented in Appendix D. An excellent review on data structures for unstructured mesh generation has been presented by Formaggia [11].

6.2.1 Boundary definition

As we limit our two dimensional mesh generation efforts to the triangulation of one closed polygonal domain with possible multiple holes a related boundary definition model can be constructed. Any closed boundary (outer boundary or hole boundary) can be represented by a closed oriented list of node indices. Some programmers would immediately suggest a linklist structure for such a case. In the implementation here, a straightforward integer vector is used and it relies on the *modulo* (or *rest*) operator to provide the connection between the end and beginning of the vector. A structure called `EDGE` that has two members is defined:

- The integer vector of node indices which define a closed boundary, and
- The number of elements in the integer vector.

The boundary of a domain consists therefore of one outer boundary (oriented in anti-clockwise fashion) and a series of inner boundaries (oriented clockwise). Maintaining this principle of orienting the boundary segments guarantees that when *walking* along the boundary the inside of the domain can always be found on the left hand side. This is similar to the input data required for the advancing front method as described in section 4.3.

A structure called `BND` is defined that has three members:

- An `EDGE` structure to represent the outer boundary of the mesh domain.
- A vector of `EDGE` structures to represent all the inner boundaries of holes inside the mesh domain.
- The number of inner boundaries which are defined in the vector of `EDGE` structures.

6.2.2 Triangle matrix

The implementation of the 2D Delaunay triangulation procedure, presented here, is built around an all-inclusive element matrix. For every triangle in the triangulation all information related to the definition of the triangle and its relation with its neighbours will be stored in the same row of the matrix. An important requirement is for the procedure to ensure that all information stored in the columns remains valid and correct at any time during the triangulation and refinement procedures.

As before, all triangles are defined anti-clockwise from a starting corner node. An order can therefore be assigned to the sides. Adopting a conventional C programming and a zero-based indexing system, the side with number `s` will be defined as the side which connects the node with index `s` to the node with index `(s+1)%3`. For example, side 2 of the triangle connects the last node (i.e. node 2) with the first node (i.e. node 0). The triangle matrix contains the following information:

- **Columns 0, 1, 2:**
  The numbers of the three corner nodes that define the triangle.

- **Columns 3, 4, 5:**
  The numbers of the adjacent triangles along the three sides of the triangle. If there is no adjacent triangle because the element lies on a boundary then a value of `-1` is used.

- **Columns 6, 7, 8:**
  The number of the side that the adjacent triangle along each of the three sides shares with this side. This information helps in determining the orientation of neighbouring triangles towards each other. When there is no neighbour along a particular side because the side is on the boundary (inner or outer) and the value in that respective column is undefined.

- **Column 9:**
  A flag (`+1/0/-1`) that is used to indicate whether the triangle has been examined to see if it is inside (`+1`) or outside (`-1`) the domain. The value `0` indicates that the triangle has not been examined. Outside triangles are part of the triangulation in order to make it convex. The flag however makes sure that these triangles will not be refined.

- **Column 10:**
  A flag (`0/1`) used to indicate whether the triangle satisfies the mesh size criteria and does not have to be refined further.

An example mesh is shown in Figure 6.4. The domain of the mesh is shaded gray. The triangulation covers the convex hull of the domain. The numbers of the triangles are shown inside a small triangle symbol. The indicators of the three sides are shown along the sides in a small square. The triangle matrix for this example is shown in Table 6.1.

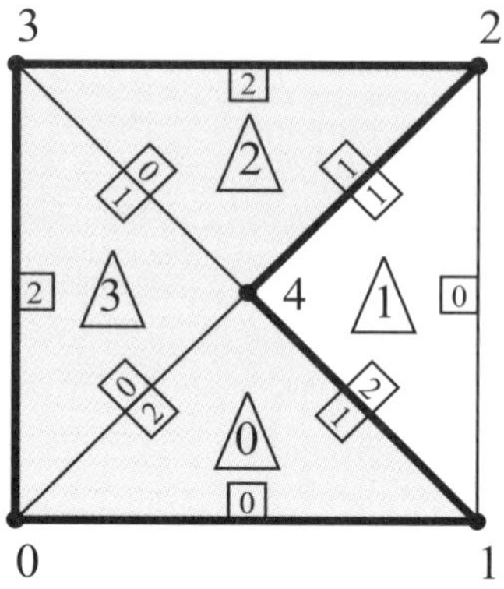

Figure 6.4: An example mesh domain and triangulation.

	0	1	2	3	4	5	6	7	8	9	10
0	0	1	4	−1	1	3	?	2	0	+1	1
1	1	2	4	−1	2	0	?	1	1	−1	1
2	3	4	2	3	1	−1	1	1	?	+1	1
3	0	4	3	0	2	−1	2	0	?	+1	1

Table 6.1: Triangle matrix for the example mesh in Figure 6.4.

As the counting of the node numbers and triangle numbers starts from 0 instead of 1, adjacent triangles can easily be accessed by taking the value of column 3, 4 or 5 as a row index. The adjacency information is also ordered in groups of three so e.g. the column index of the adjacent triangle number is the side number plus 3.

6.2.3 Nodal information

The triangle matrix has its information accessible only through the element number. Therefore the triangles which share a given node number cannot be retrieved without searching. To eliminate all searching procedures of this kind a *triangle per node* list is created. In the implementation presented here, this is a matrix called `tpnl`. It has a row of variable length for each node. The rows will be as long as necessary to store the indices of the triangles that share that node.

It is also important to know how many triangles share a particular node. This number has to be stored as well. Rather than creating another variable (vector) for this purpose the number of triangles is stored in the first column of the variable length vectors of the `tpnl` matrix. Each row has therefore a length of one plus the number of triangles that share this node.

Whenever the definition of a triangle (its three corner nodes) is changed during the triangulation procedure, the `tpnl` table has to be updated. A series of related procedures take care of the registration and removal of new and modified triangles.

6.2.4 Edge based data structure

In the element-by-element refinement method (defined in Section 1.3.5) each element from the coarse mesh will be refined individually. Boundary refinement nodes created on the edges of each coarse element might be duplicated if the edge is shared with other coarse elements. There are two approaches to avoid the definition of duplicate nodes in the mesh:

- In the first approach the boundary refinement nodes are generated independently when refining each individual coarse element. It must however be guaranteed that the location of these nodes matches the location of the boundary refinement nodes that are created along the common edge when the neighbouring triangle is being refined. This can be guaranteed by using exactly the same boundary refinement procedures with exactly the same parameters. Then there must be a merging procedure, in which the duplicate nodes are eliminated by comparing coordinates and renumbering all triangle definitions. This approach has not been implemented by the authors.

- In the second approach the boundary refinement procedure is carried out separately before the coarse triangles are separated for individual refinement. Each edge in the coarse triangular mesh is refined only once and no duplicate nodes are created. A separated edge based data structure has to be created, to contain the definitions of all the edges in the coarse mesh. This approach was chosen and implemented.

Using the definition of Section 6.2.1 an alternative vector of `EDGE` structures is created. For each of the coarse elements an `EDGE` structure is initialised with the three corner nodes of the element. This vector will be called the edge based data structure. It will be used in Section 6.5.1.

6.2.5 Important Variables

Apart from the mesh generation job description, mesh definition, boundary definition and `tpnl` other important variables used throughout the mesh generation procedure are:

- **The coordinate matrix:** The coordinates of all points and nodes involved in the meshing operation are stored in this array. The memory allocation strategy is based on allocating extra rows in the matrix in discrete chunks rather than one at the time. Therefore two integer variables are kept. One indicates the current number of nodes and the other indicates the number of rows allocated in the matrix.

- **Mesh parameter vector:** A vector which contains for each node in the mesh the required edge length size for the edges connected to that node.

- **The background mesh variables:** If the meshing method is background controlled (as defined in Section 3.3.2) a complete set of variables related to the background mesh and its mesh size requirements is used. This includes a separate triangle matrix, a coordinate matrix and a mesh parameter vector.

6.3 Program layout

The layout of the Delaunay triangulation part of the mesh generator will depend on which mesh generation task the user requires it to perform. The different possible mesh generation tasks are mentioned in Appendix C. Depending on the selected task the program will take one of the different paths through the flowchart. Top-level functions for each of the paths will be different but they all share the same basic core-level functions on which the Delaunay triangulation kernel is built. Paths may join at a later stage in the processing line e.g. when it comes to preparing the generated data for output in a suitable format.

Figure 6.5 shows the top-level flowchart of the Delaunay triangulation part of the mesh generator. A short description of each of the nine top-level steps is given. Several of the steps are decomposed into lower-level flowcharts in subsequent figures and descriptions.

∘ 1. **parse the mesh into internal data structures**
In section 6.4.2 it will be explained how the high-level mesh definition data structure is parsed into a series of separate data structures. The separate steps required in this process are shown in the flowchart in Figure 6.6.

∘ 2. **task selection**
The value of the task descriptor (`EBE`, `REMESH` or `REFINE`) determines which path is selected in the mesh generation algorithm.

∘ 3. **refine edges**
The edge refinement procedure is carried out on the edge based data structure created in step 1.14 . This procedure will be described in detail in Section 6.5.1.

∘ 4. **refine element-by-element**
The refinement strategy which will be shown in Section 6.7 is applied to every element of the coarse mesh.

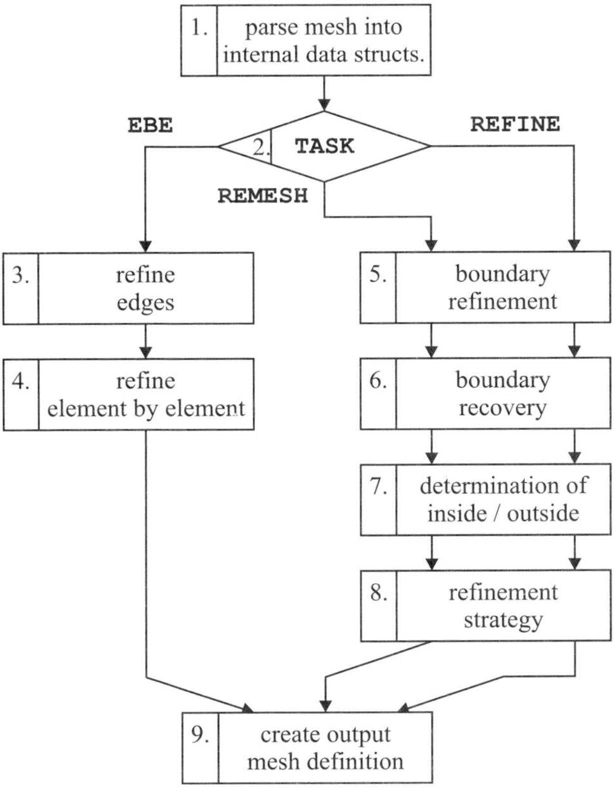

Figure 6.5: Top-level flowchart.

- ∘ 5. **boundary refinement**
  The algorithm for the boundary refinement procedure is shown as a flowchart in Figure 6.12. In Section 6.5 the fine details of the algorithm will be explained.

- ∘ 6. **boundary recovery**
  For a non-convex mesh domain a boundary recovery procedure is required. This procedure is shown in the flowchart in Figure 6.20. The detailed description follows in Section 6.8.1.

- ∘ 7. **determination of inside/outside**
  Once all boundary segments are recovered all triangles in the current triangulation can be marked as either inside or outside the mesh domain. The procedure that determines this property is described in Section 6.8.2.

- ∘ 8. **refinement strategy**
  This is the core of the mesh generation program. The refinement strategy se-

lected in the Delaunay option descriptor is applied only to the triangles inside the domain as marked in Step 7. Refining oniy the internal triangles and constraining the algorithm not to modify the boundary edges ensures that there is no need for another boundary recovery step later. The different refinement options and strategies will be discussed in Section 6.7.

○ 9. **create output mesh definition**
The last step of the entire mesh generation procedure is the output of the generated fine mesh. The coordinates of the freshly generated refinement nodes and the definitions of the new triangles are copied into a mesh definition data structure. All obsolete information is discarded and allocated memory is freed up. The output mesh definition data is written to an output file of which the name is stored in the mesh generation job description.

6.4 Pre-processing tasks

In order to accommodate a wide range of tasks from initial mesh generation to adaptive refinement a significant number of pre-processing tasks have to be undertaken. The mesh generation task must be reduced to those elementary building blocks which are common for all tasks. The pre-processing stage of the mesh generation program therefore consists of procedures such as reading in the input files, checking their validity, converting them into the internal data structures and recreating all missing information. Figure 6.6 shows the separate steps required as part of the pre-processing:

○ 1.1 **parse the mesh definition into the triangle matrix**
The defining node numbers of all triangles in the mesh definition data structure are copied into the triangle matrix for easier handling during the mesh generation procedures.

○ 1.2 **check that all triangles are ACW**
To maintain consistent definitions of triangular elements during the process of mesh generation all triangles of the coarse mesh must be defined in an anticlockwise order. This procedure checks this requirement.

○ 1.3 **retrieve the adjacency information for triangle matrix**
The matrix that contains the definitions of the triangles also contains information concerning the adjacency between the triangles. As this information is required during the mesh generation process it must be retrieved or reconstructed for the input coarse mesh.

○ 1.4 **task selection**
Once the loaded mesh is converted into the internal triangle matrix some extra operations are carried out. These depend on the task description that has to be completed. Possible task descriptors are `EBE`, `REFINE` and `REMESH` as defined in Appendix C.

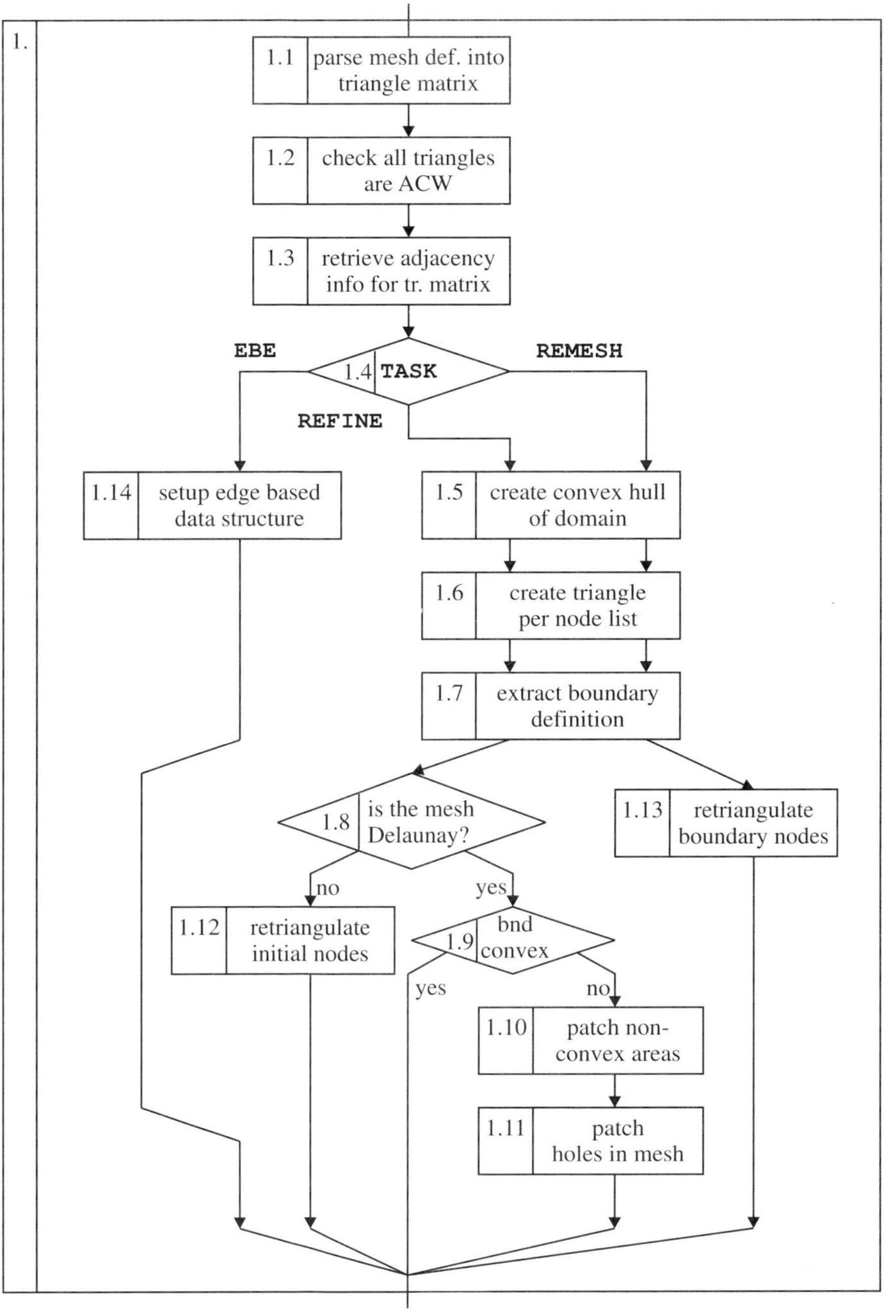

Figure 6.6: Flowchart for: 1. parse mesh into internal data structures.

- ○ 1.5 **create the convex hull**
  The convex hull of the domain represented by the coarse mesh is determined. The methods to determine which points from the coarse mesh define the convex hull and in what order will be explained in Section 6.4.3.

- ○ 1.6 **create the triangle-per-node list**
  The triangle-per-node list as defined in Section 6.2.3 is initialized and filled for the nodes and triangles of the coarse mesh on the basis of the information stored in the recently created triangle matrix.

- ○ 1.7 **extract the boundary definition**
  Using the information from the triangle matrix and the triangle-per-node list the boundary definition of the coarse mesh can be extracted in the format defined in Section 6.2.1. The procedure to build the boundary definition will be described in Section 6.4.4.

- ○ 1.8 **is the mesh Delaunay?**
  For the task `REFINE` a checking procedure has to be carried out. Depending on whether the coarse mesh satisfies the criteria for a valid Delaunay mesh different actions are taken. The procedure to check the Delaunay properties of a triangulation are detailed in Section 6.4.1.

- ○ 1.9 **is the boundary convex?**
  If the Delaunay property checking confirms that no violation of the Delaunay criterion is detected the loaded coarse mesh can be used for further refinement on condition that the triangulation covers the convex hull of the domain. A test is therefore carried out to identify any non-convex areas in the mesh domain. The implementation of this test will be discussed in Section 6.4.5.1.

- ○ 1.10 **patch the non-convex areas**
  If the mesh domain is not convex the non-convex areas can be triangulated separately and patched into the existing triangulation of the mesh domain. The patching procedure will be described in Section 6.4.5.2.

- ○ 1.11 **patch the holes in the mesh**
  If any holes exist in the mesh domain the individual triangulation of these holes can be performed and merged with the existing triangulation of the mesh domain. Section 6.4.5.4 will contain the procedure that describes the peculiarities of this merging operation.

- ○ 1.12 **retriangulate the initial nodes**
  However if the Delaunay property checking identifies that the coarse mesh is not the Delaunay triangulation of the initial nodes then the only possibility to continue the refinement path is to retriangulate the initial nodes. Using the Delaunay procedures which will be implemented in Section 6.6 the Delaunay triangulation of the convex hull of the initial nodes is created.

○ 1.13 **retriangulate the boundary nodes**
The purpose of remeshing is to keep only the boundary definition and to discard any internal nodes in the coarse mesh. The triangle matrix that was created from the triangles defined in the coarse mesh must therefore be replaced by the result of a triangulation procedure carried out on the boundary nodes only. The creation of a retriangulation of a series of nodes will be described in Section 6.6.5. The boundary definition data, created in step 1.7 is used to identify which nodes belong to the boundary and should be included in the retriangulation.

○ 1.14 **setup edge based data structure**
The element by element refinement method treats each element separately as an individual mesh generation problem. For reasons mentioned in Section 6.2.4 the edge based data structure has to be created from the information stored in the triangle matrix of the coarse mesh.

Each of these steps is now discussed in greater detail.

6.4.1 Delaunay validity checking

The Delaunay triangulation kernel requires that at any time a valid Delaunay triangulation is preserved. If a triangulation is performed afresh then the incremental point insertion algorithm guarantees that the Delaunay integrity is preserved whenever a new point is inserted. However to refine a given triangulation or to continue with an aborted refinement process the triangulation must be a Delaunay triangulation. This must be checked first. Checking the Delaunay validity is unfortunately a very time-consuming task.

In the simplest version without further optimizations all nodes have to be checked against all triangles for the in-circle criterion. A fast implementation of the in-circle criterion was suggested by Guibas and Stolfi [15]. As shown in Figure 6.7 a node D belongs to the circumcircle of triangle ABC if

$$\begin{vmatrix} x_A & y_A & x_A^2 + y_A^2 & 1 \\ x_B & y_B & x_B^2 + y_B^2 & 1 \\ x_C & y_C & x_C^2 + y_C^2 & 1 \\ x_D & y_D & x_D^2 + y_D^2 & 1 \end{vmatrix} > 0$$

The determinant of this 4 by 4 matrix is easily calculated as the sum of four determinants of 3 by 3 matrices as the last column of the matrix contains only unity values.

It is obvious that this checking on large meshes must be avoided at all cost as it will require considerable computational time.

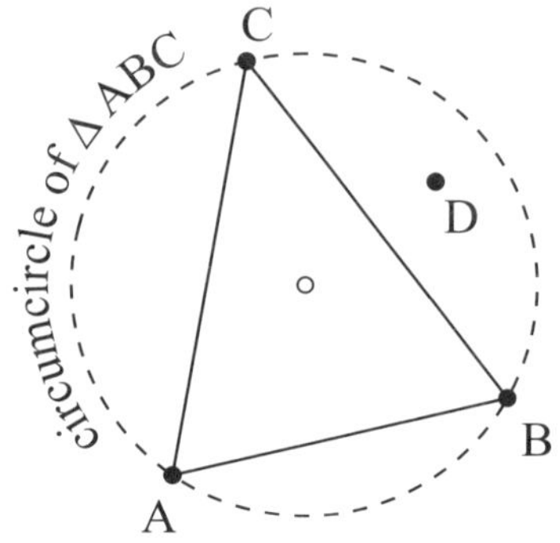

Figure 6.7: Illustration of the incircle criterion.

6.4.2 Parsing coarse meshes

To exploit the power of the all-in-one element matrix, coarse meshes have to be converted to this format. The external mesh definition file format only gives definitions of each triangle i.e. three corner node numbers per triangle. The adjacency information has to be retrieved to fill in the remaining columns of the triangle matrix.

The adjacency information can be retrieved using the *triangle per node* list. The neighbour of a given triangle along a given side can be found by comparing the `tpnl` entries of the two nodes that define the side. The `tpnl` entries (i.e. the rows of the `tpnl` matrix) are lists of all triangles that have a particular node as one of its corner nodes. The intersection (common triangle numbers) of the two triangle lists may be one or two triangles.

If only one triangle is returned it must be the given triangle. If two triangles are returned one of them must be the given triangle and the other triangle must be the neighbour. Once one neighbour relation is found the inverse neighbour relationship can be set thus reducing the number of search operations by a factor of two.

6.4.3 Creating the convex hull

The convex hull of a cloud of points in two dimensions is the smallest convex polygon made out of a subset that contains all points. There are several algorithms for retrieving the convex hull. These algorithms include: *divide-and-conquer* [8], *quickhull* [1] and *giftwrap* [5]. Most of them rely on the fact that the position of all points is known in advance and they are not very efficient to redo when a new point has to be added. An incremental algorithm is therefore advisable. Throughout the entire meshing operation the convex hull of the domain will be maintained. Whenever new points are created in the refinement procedure the convex hull will be updated.

Every time a new point is added a check is performed to see if it falls inside or outside the current convex hull. If the point falls inside no action is required. If the point falls outside, the hull has to be extended. Because the hull is convex a point falls inside it if it lies on the left hand side of each segment of the hull assuming that the hull segments are numbered anti-clockwise.

By looping through all segments of the hull it is easily checked if the point in question lies on the left or right hand side of the segments. The point lies on the left hand side of a segment if the triangle made by the first node of the segment, the second node of the segment and the new point (in that order) is an anti-clockwise triangle. The *clockwise* test can be implemented as a determinant calculation [15]. A triangle ABC is clockwise if

$$\begin{vmatrix} x_A & y_A & 1 \\ x_B & y_B & 1 \\ x_C & y_C & 1 \end{vmatrix} > 0 \tag{6.1}$$

Once it has been identified that a point lies on the right hand side of a hull segment the algorithm keeps on looping through the following segments until the first segment has been encountered for which the point lies on the left hand side. The series of segments on whose right hand side the point falls have to be replaced by two new segments: i.e. the connections with the new point as shown in Figure 6.8. Depending on how many segments of the old hull become invalid the hull vector may have to be increased in length or shortened.

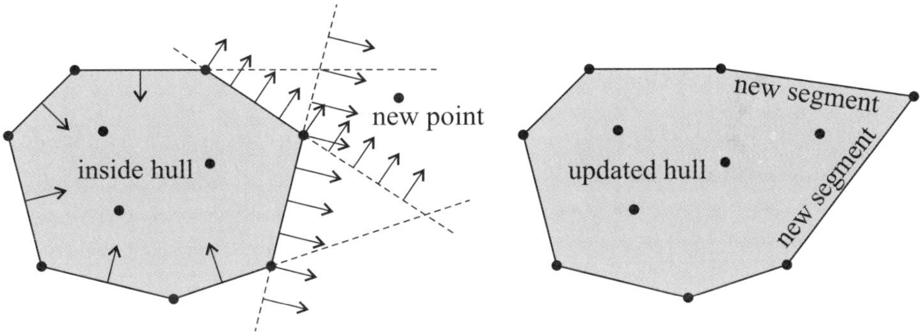

Figure 6.8: Updating the convex hull.

6.4.4 Boundary extraction

Next the extraction of the boundary definition of a given mesh into the format specified in Section 6.2.1. All segments of either the outer boundary or the hole boundaries of a given mesh must belong to triangles with an incomplete set of neighbours (i.e.

fewer than three neighbours). Therefore as a prerequisite to this procedure the triangle matrix (complete with all adjacency information) is required. All triangles which have their complete set of neighbours are marked in a special vector.

Topologically it is not possible to distinguish between the outer boundary and any of the boundaries of cutouts or holes unless one node that definitely belongs to the outer boundary is known. In the procedures implemented here, that information is taken from the convex hull vector. Every node of the convex hull must belong to the outer boundary. The first node of the convex hull vector is therefore taken as the starting point of the outer boundary.

The algorithm will now walk from the starting point along the outer boundary until the starting point is reached again and the boundary is closed.

6.4.4.1 Walking procedure to retrieve a boundary

Once again the combination of the adjacency data (stored in the triangle matrix) and the `tpnl` table gives enough information to step from triangle to triangle while following a boundary. The procedure starts with the initial node and the initial node is looked up in the `tpnl` table. All triangles recorded in the the initial node's entry in the `tpnl` table are examined. A triangle is sought that has no neighbour on the side which commences at the initial node. If there is no neighbour it means that this triangle must lie on the boundary and that the next segment of the boundary is therefore a segment of the triangle which contains the initial node. The enforced anti-clockwise orientation of both the triangle and the boundary definition dictates that it must be the segment that commences with the initial node. The first segment of the boundary is now recovered. To proceed with finding the rest of the boundary the initial node is replaced with the end node of the segment and the procedure is repeated. The triangle is also marked as examined.

6.4.4.2 Detecting hole boundaries

Once the outer boundary is retrieved all remaining unmarked triangles that have an incomplete set of neighbours must then belong to the boundary of a hole. The walking procedure can then be restarted from one of these unmarked triangles until the hole boundary is recovered. The procedure keeps repeating as long as there are unmarked triangles remaining.

6.4.5 Dealing with non-convex coarse meshes

If however a mesh for a non-convex domain is required, a series of triangles must have been generated belonging to the convex hull of the domain but not belonging to the actual domain itself, because the DT kernel can only handle convex domains.

The triangles in the non-convex areas were then removed from the final mesh before output. If later on it is decided to refine the mesh even further, these non-convex areas have to be patched in again.

The engineering solution to this problem is to isolate the areas that have to be patched in and triangulate them as if they were a separate meshing problem. Subproblems of this approach are:

- The recognition of these non-convex areas and their isolation;
- the triangulation of these non-convex areas; and
- the merger of the triangulated patches back with the main input triangulation without corrupting the adjacency data.

6.4.5.1 Recognition of the non-convex areas

A non-convex area of a domain can be recognized from the internal data structures when comparing the boundary data with the convex hull data. It is important to remember that both the outer boundary description and the convex hull are stored as vectors of node numbers. When the domain is convex, the convex hull and the outer boundary of the domain are identical apart from a possible offset of the starting points since it was assured that orientations of the node number vectors are anti-clockwise. However for a non-convex domain discrepancies between the convex hull and the outer boundary will arise since the outer boundary will have more node numbers in it than the convex hull vector. This provides a clue as to how to isolate the non-convex areas.

First the offset between the outer boundary vector and convex hull vector is determined. As every node of the hull, by definition, belongs to the outer boundary it is easy to take the first node in the convex hull vector and search through the outer boundary vector until that node number is located. Stepping through both vectors will produce the same numbers as long as the nodes represent a convex region. As soon as different numbers are encountered then this is an indication of the start of a non-convex region.

Figure 6.9 shows by example how the procedure works to determine the boundary of the non-convex area (the *patch*). Once a discrepancy between the convex hull and the outer boundary has been determined, the boundary of the non-convex area is constructed using the two consecutive nodes from the convex hull that are involved and the excess outer boundary nodes in reverse order.

6.4.5.2 Triangulating non-convex areas

This triangulation should not pose any problems as it uses the same DT procedures which will be applied at a later stage and are described in section 6.6. The only problem that might arise is when the isolated non-convex area is non-convex. The triangulation procedure will then generate a triangulation of the convex hull of the

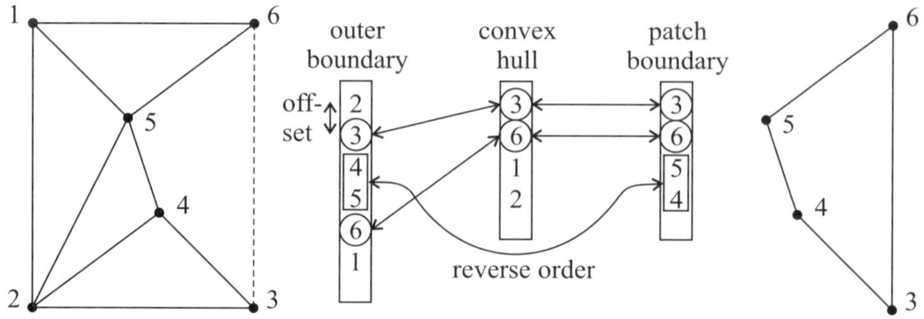

Figure 6.9: Recognizing non-convex areas.

patch area. Some of the triangles will have to be removed before merging can take place. However this is not an exceptional procedure as it will be applied on the scale of the full mesh at the end of the full meshing procedure.

Figure 6.10 shows an example of a non-convex area of the domain that is non-convex itself. The steps of triangulation and removal of excess triangles are also shown.

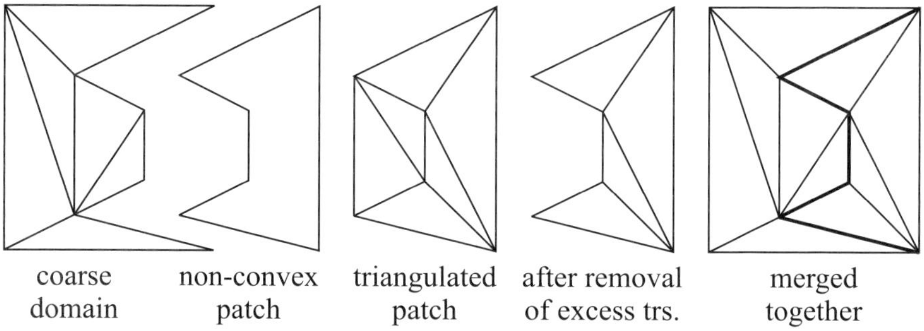

Figure 6.10: Triangulating non-convex areas.

6.4.5.3 Merging triangulated patch areas

The triangle matrix of the triangulated patch area is easily appended to the triangle matrix of the loaded coarse mesh. The triangle definitions remain valid as the correct node numbers were used since they came from the convex hull and boundary vectors. However, the numbers these new triangles will receive are different. When appending each new triangle, a new number is assigned starting from the last number used in the coarse mesh. This assignment will corrupt the internal adjacency information between the triangles of the patch area. Also if some of the triangles of the patch area were removed for the reasons explained above, simply adding the previous total number of triangles to each new triangle number will corrupt the data. One way to solve this is to use a translation table.

This table is a simple integer vector with as many entries as there are new triangles in the triangulated patch (before removal of excess triangles as a result of the non-convexity of the patch).

The process of building that table is demonstrated in Figure 6.11. Each entry in the vector will be filled with the new number that the triangle will receive when it is copied onto the end of the main triangle matrix. Then the numbers for the excess triangles are removed and the higher numbers are decremented. The table is then used to update the internal adjacency data. In the example given in Figure 6.11 the numbers of the neighbouring triangles of triangle 1 of the patch are 3 and 2 which will be translated into -1 (i.e. no neighbour) and 8.

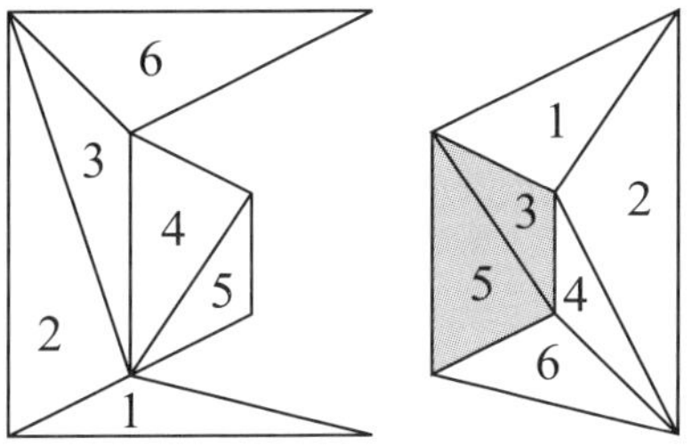

original △ number	+6	remove △3	translated △ number
1	7	7	7
2	8	8	8
3	9	-1	-1
4	10	9	9
5	11	10	-1
6	12	11	10

Figure 6.11: Creating a translation table for a triangulated patch area.

The outside adjacency information (no longer inside the patch but over the borders of the patch) needs some extra attention. The link with the main domain over the shared portion of the outer boundary has to be restored. Fortunately the shared portion of the boundary is known. By means of the `tpnl` list a procedure identical to the one used in Section 6.4.2 may be used to identify triangles that lie on both sides of the common boundary segments. Once pairs of neighbouring triangles are identified their relative adjacency can be corrected.

The only remaining task is to update the `tpnl` list. All that has to be done is to register all new triangles in the `tpnl` using a registration function.

6.4.5.4 Dealing with holes

The same approach may be applied to holes that have to be patched in. Finding the boundary is even easier as it can be taken directly from the boundary description of the mesh. However care needs to be taken concerning the orientation of the boundary vector. In the boundary definition of the mesh the node numbers that form a hole are stored in a clockwise fashion. The boundary nodes for an area that has to be meshed are stored anti-clockwise. The order therefore has to be reversed.

Other concerns such as dealing with non-convex holes and how to merge a triangulated hole back into the mesh are identical to the procedures described above for non-convex areas of the domain.

6.5 Boundary refinement

Boundary refinement comprises all tasks required to create a boundary approximation that satisfies the required mesh size criteria. This means that the length of each segment on the boundary is not greater than the size criterion that can be derived for the segment's two vertices. In general a coarse mesh which is loaded at the start of the meshing operation will have boundary segments far exceeding the required length criteria.

The flowchart in Figure 6.12 outlines the steps that are to be taken for normal boundary refinement. The different control methods to determine mesh size criteria are discussed in Chapter 3. The actual procedures for creating new refinement nodes on straight and curved boundary segments are described in Appendix A.

In the case of element-by-element based mesh refinement there are some complications which are discussed in the next section.

6.5.1 Refining the edge based data structure

The edge based data structure was set up to avoid refining sides of the triangles twice when undertaking element-by-element refinement. The three sides of each triangle are refined independently. First the triangle is checked to determine if a neighbouring triangle adjacent to the side exists. If there is no neighbouring triangle then the side can be refined as usual using the segment refinement method described in Appendix A. If a neighbour does exist the following procedure is executed:

- If the neighbouring triangle has a higher number than the current triangle then the common side has not yet been refined, as the neighbouring triangle has not been dealt with. The segment should be refined with the segment refinement method described in Appendix A.
- However if the neighbouring triangle has a lower number it has already been

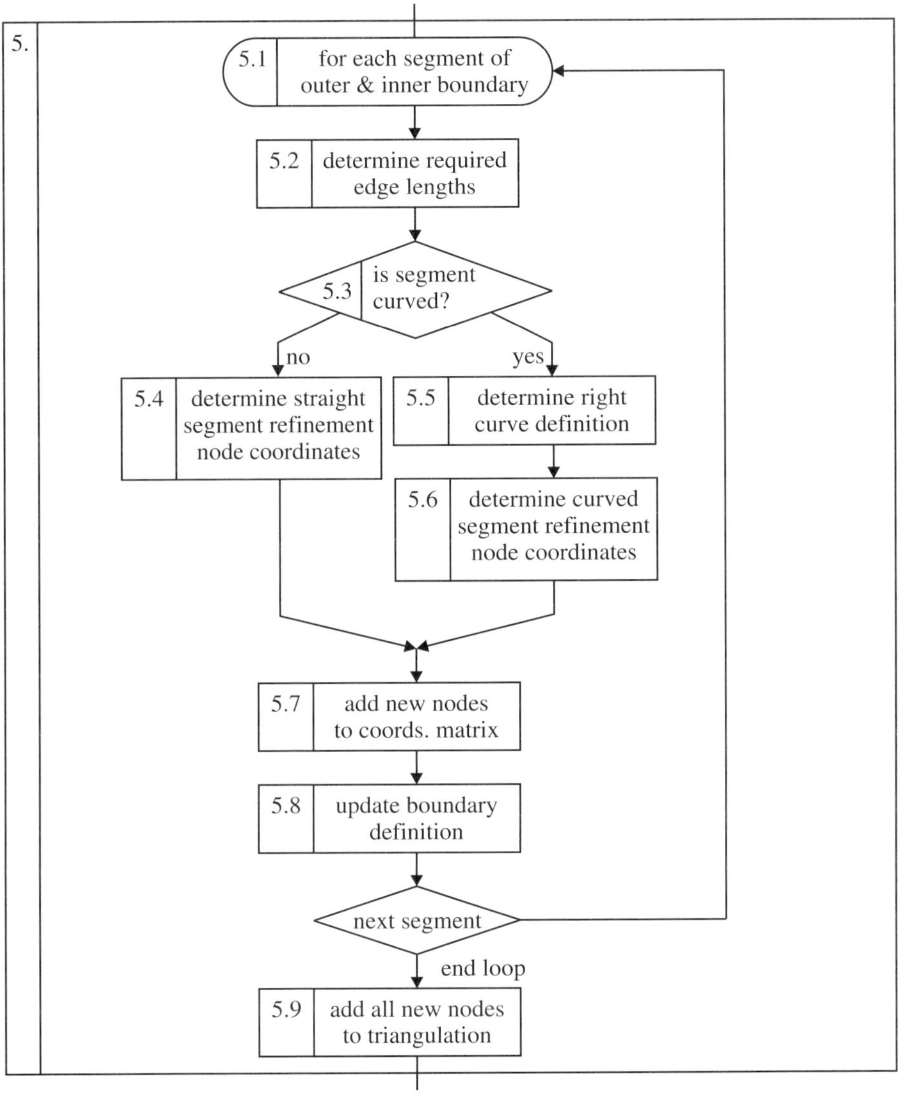

Figure 6.12: Flowchart for: 5. boundary refinement.

visited and the common side has been refined. The following actions should be taken.

- The location of the two nodes which form the segment is looked up in the `EDGE` structure of the neighbouring triangle.
- If the two nodes are consecutive in the neighbour's edge definition the

segment has been refined when the neighbouring triangle was dealt with and no refinement nodes were required as the segment already met the mesh size criteria. No further action is necessary and the procedure for this edge finishes here.

- ◦ If the nodes are not consecutive a series of boundary refinement nodes were created when the neighbouring triangle was dealt with. The numbers of those refinement nodes are copied into the edge definition of the current triangle in reverse order to create a proper match.

6.6 Delaunay procedures

The insertion of a new point into an existing triangulation is the main idea of the incremental DT algorithm. Three steps always have to be taken:

- Locating the current triangle that the new point belongs to;
- Splitting up this triangle and reconnecting the new point; and
- An edge flipping procedure: restoring the Delaunay integrity in the triangulation.

Variations on this theme are: connecting a new point which falls outside the current domain to the triangulation. The algorithms behind these steps will now be discussed.

6.6.1 Location search

The task is to identify a triangle in a triangulation to which a point with given coordinates belongs. This interesting problem was solved by Borouchaki *et al* [3] who suggested a walking algorithm. The algorithm requires a starting triangle from where the walk towards the final triangle will start.

Two variations were suggested to determine the side that will lead to the next element in the walking order:

- An approach based on intersection: Determine the centroid of the current triangle and find out which of the three sides is crossed by the connecting line between the point and the centroid.
- An approach based on visibility: Take the first side that is not visible to the point: i.e. the first side for which the point lies on the right hand side of the line through the side.

Input: Starting triangle S; Point P
Output: Triangle T enclosing point P

```
(1)   c := S
(2)   if Point P is inside triangle c
(3)       return Triangle c
(4)   else
(5)       f := side through which to advance to the next triangle
(6)       if No neighbour at face f
(7)           return error, point is unreachable
(8)       else
(9)           c := neighbour triangle
(10)          goto Line 2
```

Algorithm 6.1: Walking algorithm.

Borouchaki *et al* [3] demonstrated that the approach based on visibility may have drawbacks as looping situations may occur whereby the the point may never be reached. This is why the implementation proposed in this book adopts the intersection approach although its algorithmic complexity might be greater. Figure 6.13 illustrates the algorithm with the approach based on intersection. The only computational question is

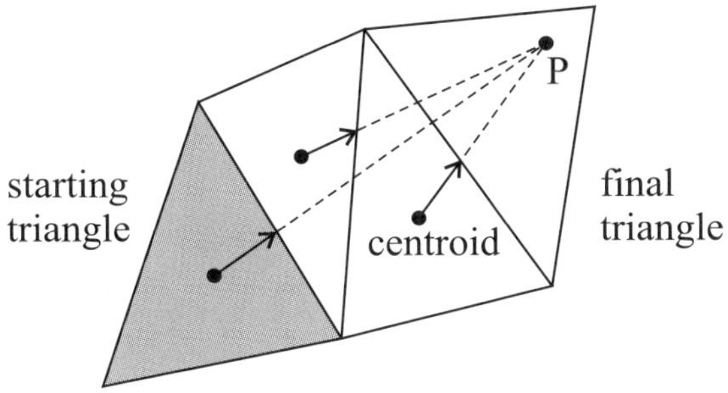

Figure 6.13: Demonstration of the walking algorithm.

to determine whether a point lies inside a triangle. Some programmers resolve this question by taking the sum of the areas of the three triangles that are obtained by connecting the point to any of the three sides of the triangle. If the sum equals the area of the triangle the point lies inside. If the point lies outside the triangle the sum will always be greater than the area of the triangle. This method requires the computation of the areas of four triangles.

An easier solution might be suggested that relies on the fact that all triangles are originally numbered anti-clockwise. The three triangles which are constructed by taking two nodes from the side of the triangle (in that order) followed by the point should all three be anti-clockwise if the point lies inside the triangle. Figure 6.14 demonstrates the principle. The implementation of the clockwise test has already been explained in Section 6.4.3.

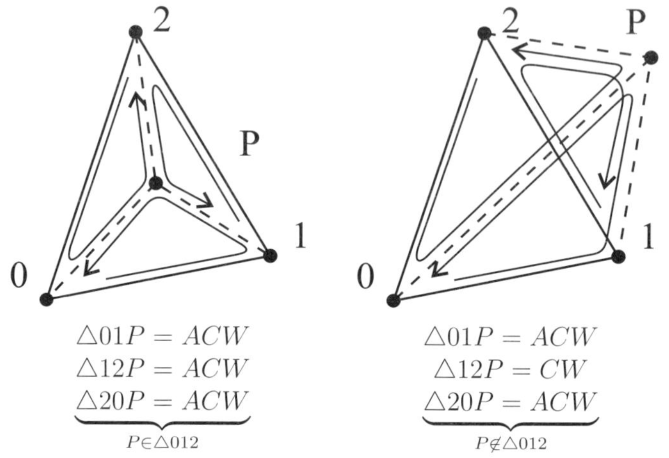

Figure 6.14: Testing if a point lies inside a triangle.

This simple algorithm can be modified to accommodate two special cases:

- If the point that has to be located lies on one of the sides of a triangle it would be appropriate for the location search procedure to reveal this and hence indicate this special case. If the clockwise test is implemented using Equation 6.1 the determinant will become zero if the point lies on one of the sides. This can be detected and the triangle number plus a side indicator can be returned.

  As these calculations are carried out in double precision arithmetic a value of 0.0 will rarely be obtained. To give the program a certain robustness a threshold value is introduced. Once the absolute value of the determinant of Equation 6.1 is smaller than the threshold it is assumed that the node lies on the side corresponding with that clockwise test.

- If the point happens to coincide with any of the current points in the triangulation (i.e. a duplicate point was created) the location search routine should return the triangle and the node number of the duplicate. When two determinants out of three return a value below the accuracy of the threshold it is safe to say that the point is close enough to one of the corner nodes of the current triangle.

The implemented location search routine has therefore three possible return values:

- $+1$: the search procedure has located a triangle and the point lies completely inside it or lies on one of its sides. This distinction is made by the value of the second return value `onside` which can be:
  - -1: the point lies completely inside the located triangle
  - 0, 1 or 2: the point lies on one of the three sides of the located triangle
- 0: the point lies outside the current triangulation because during the walking procedure a boundary was encountered before the point was reached.
- -1: a triangle was located but the point is a duplicate of one of its corner nodes. The second return value `onside` in this case contains 0, 1 or 2 to indicate which node of the triangle is the duplicate.

To minimize the number of steps in the walking procedure a good guess must be made of which starting triangle is in close proximity to the new point. If consecutive points which must be located are known to be close to each other and triangle numbers are geographically sorted the result of a previous search operation can be used as the starting point for the next operation.

6.6.2 Reconnection

Once the refinement point has been located, the new point has to be inserted into the triangulation. If the point was a duplicate it can be ignored. In all other cases one of the following reconnection patterns is used. In the following figures and tables these patterns will be presented. The reconnection pattern is a representation in tabular form demonstrating how to modify the element matrix that contains the definition of the mesh to include the triangles associated with the insertion of the new point.

By always using these reconnection patterns it is guaranteed that adjacency information stored in the element matrix is not corrupted. A reconnection pattern consists of a figure (showing the initial triangle, the layout of its neighbours and orientation of its sides) and an extract of those rows in the element matrix that are to be modified. Only columns 0 to 8 of the triangle matrix are shown as they contain the triangle definition and adjacency information. As defined in Section 6.2.2 columns 0, 1 and 2 contain the triangle definition, columns 3, 4 and 5 contain the neighbours of the triangle and columns 6, 7 and 8 contain the side numbers that are shared with the neighbours.

All the patterns share the following conventions:

- The new node that is to be inserted will always become node 0 of all the new triangles which are created.
- The number of the triangle that will be deleted (as a result of reconnection) is reused by one of the new triangles which are created. All other new triangles will be given entries at the end of the triangle matrix.

The convention used in the description of the pattern is that n_i always stands for a node number, t_i stands for a triangle number and s_i stands for a side indication, i.e. a value of 0, 1 or 2 to indicate the first, second or third side of the triangle. A star ($\star$) in the table indicates that the value is not important and is not shown in the figure. And finally when three values are enclosed by a curly bracket a cyclic rotation backwards as many times as shown by the subscript is required to obtain the correct position of the values in the arrays. This cyclic rotation is required for triangular elements where their definition depends on some other value in the table which determine the actual order of the nodes of the triangles. For example the definition of triangle $\{\star\ n_0\ n_1\}_{s_a}$ means that the triangle is defined by three nodes: an unknown node, n_0 and n_1 in the specified order, but the value of s_a determines whether n_0, n_1 or the unknown node is the first node. If the value of s_a is 0 the unknown node is the first node, n_0 is the second node and n_1 is the third node of the triangle. When the value of s_a is 1 then node n_0 is the first node, n_1 is the second node and the unknown node is the third node of the triangle. The order of the nodes for s_a equal two can be determined similarly.

6.6.2.1 Inserting a node that lies within a triangle

Figure 6.15 shows the reconnection pattern to be used when the new point lies completely within the triangle that was the result of the location search. Table 6.2 shows the modifications that are to be made in the element matrix.

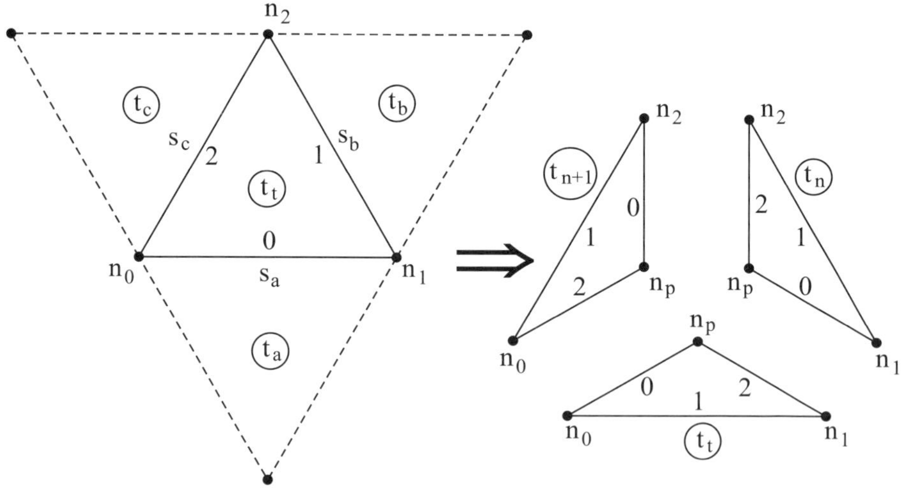

Figure 6.15: Standard reconnection pattern when inserting a node that lies completely within a triangle.

As seen from the left hand side of Figure 6.15 triangle t_t is defined by nodes n_0, n_1 and n_2. It has on its sides 0, 1 and 2, neighbours t_a, t_b and t_c which have triangle t_t on their

Triangle matrix before insertion

		0	1	2	3	4	5	6	7	8
...										
t_t	$\longrightarrow$	n_0	n_1	n_2	t_a	t_b	t_c	s_a	s_b	s_c
...										
t_a	$\longrightarrow$	$\{\ \star$	n_1	$n_0\ \}_{s_a}$	$\{\ \star$	t_t	$\star\ \}_{s_a}$	$\{\ \star$	0	$\star\ \}_{s_a}$
t_b	$\longrightarrow$	$\{\ \star$	n_2	$n_1\ \}_{s_b}$	$\{\ \star$	t_t	$\star\ \}_{s_b}$	$\{\ \star$	1	$\star\ \}_{s_b}$
t_c	$\longrightarrow$	$\{\ \star$	n_0	$n_2\ \}_{s_c}$	$\{\ \star$	t_t	$\star\ \}_{s_c}$	$\{\ \star$	2	$\star\ \}_{s_c}$
...										

Triangle matrix after insertion

		0	1	2	3	4	5	6	7	8
...										
t_t	$\longrightarrow$	n_p	n_0	n_1	t_{n+1}	t_a	t_n	2	s_a	0
...										
t_a	$\longrightarrow$	$\{\ \star$	n_1	$n_0\ \}_{s_a}$	$\{\ \star$	t_t	$\star\ \}_{s_a}$	$\{\ \star$	1	$\star\ \}_{s_a}$
t_b	$\longrightarrow$	$\{\ \star$	n_2	$n_1\ \}_{s_b}$	$\{\ \star$	t_n	$\star\ \}_{s_b}$	$\{\ \star$	1	$\star\ \}_{s_b}$
t_c	$\longrightarrow$	$\{\ \star$	n_0	$n_2\ \}_{s_c}$	$\{\ \star$	t_{n+1}	$\star\ \}_{s_c}$	$\{\ \star$	1	$\star\ \}_{s_c}$
...										
t_n	$\longrightarrow$	n_p	n_1	n_2	t_t	t_b	t_{n+1}	2	s_b	0
t_{n+1}	$\longrightarrow$	n_p	n_2	n_0	t_n	t_c	t_t	2	s_c	0
...										

Table 6.2: Standard reconnection pattern when inserting a node that lies completely within a triangle.

respective sides s_a, s_b and s_c. In the top half of Table 6.2 the rows for triangle t_t and its three neighbours are shown. As the orientation of the neighbours towards triangle t_t is not known, the cyclic notation with curly brackets is used to put all information in the correct place in the rows of the element matrix.

The new node n_p is found to be lying completely within triangle t_t. Triangle t_t will therefore be split into three triangles as shown in the right hand side of Figure 6.15. These three new triangles will be t_t, t_n and t_{n+1}. As stated before the triangle number t_t will be recycled. t_n and t_{n+1} are new triangle numbers at the bottom of the element matrix.

The new triangle t_t will be defined from nodes n_p (the new node), n_0 and n_1. The neighbours of the new triangle t_t are the two other new triangles and the neighbour that the old triangle t_t had on its first side: i.e. triangle t_a. The lower half of Table 6.2 shows the row that corresponds with the new definition of triangle t_t. Triangle t_n will be created from nodes n_p, n_1 and n_2. The new node n_p is always taken as the first node for any new or redefined triangles. Triangle t_n will have as neighbours: the redefined triangle t_t, the other new triangle t_{n+1} and the second neighbour of the old t_t: i.e. triangle t_b. The row in the element matrix for new triangle t_n is show in the table.

Some modifications also have to be made in the definitions of the triangles t_a, t_b and t_c, because they no longer have the old triangle t_t as their exclusive neighbour. They now have three different neighbours: the new t_t, t_n and t_{n+1}. The modifications are shown in the updated rows for t_a, t_b and t_c, shown in the table.

6.6.2.2 Inserting a node that lies on a side of a triangle

If the location search indicates that the new point belongs to one of the sides of the triangle a different reconnection pattern is required.

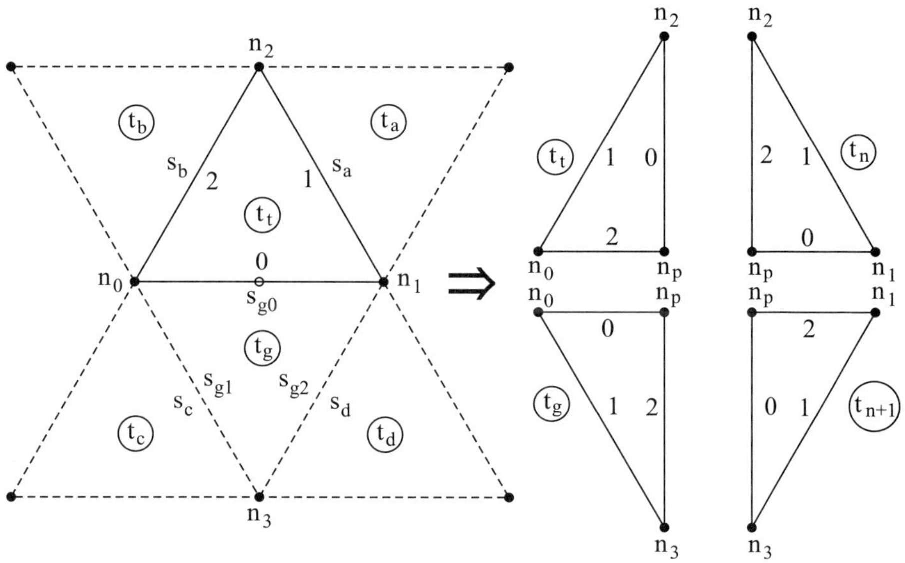

Figure 6.16: Reconnecting a node on side 0 of a triangle.

If the new node (which is given number n_p) is found on side 0 of triangle t_t, Figure 6.16 should be used in conjunction with the pattern from Table 6.3. On the left hand side of Figure 6.16 it is shown that triangle t_t has triangle t_g on its first side. Both triangle t_t and t_g will therefore be affected by the insertion of node n_p. They will both be split into two. A total of four triangles is therefore created. Two of the four will recycle the triangle numbers t_t and t_g. The other two will be given numbers at the bottom of the element matrix: i.e. t_n and t_{n+1}. The other triangles which are influenced by the splitting are the neighbours of t_t and t_g: i.e. triangles t_a, t_b, t_c and

Triangle matrix before insertion

		0	1	2	3	4	5	6	7	8
...										
t_t	$\rightarrow$	n_0	n_1	n_2	t_g	t_a	t_b	s_{g0}	s_a	s_b
t_g	$\rightarrow$	$\{\ n_3$	n_1	$n_0\ \}_{s_{g0}}$	$\{\ t_d$	t_t	$t_c\ \}_{s_{g0}}$	$\{\ s_d$	0	$s_c\ \}_{s_{g0}}$
...										
t_a	$\rightarrow$	$\{\ \star$	n_2	$n_1\ \}_{s_a}$	$\{\ \star$	t_t	$\star\ \}_{s_a}$	$\{\ \star$	1	$\star\ \}_{s_a}$
t_b	$\rightarrow$	$\{\ \star$	n_0	$n_2\ \}_{s_b}$	$\{\ \star$	t_t	$\star\ \}_{s_b}$	$\{\ \star$	2	$\star\ \}_{s_b}$
t_c	$\rightarrow$	$\{\ \star$	n_3	$n_0\ \}_{s_c}$	$\{\ \star$	t_g	$\star\ \}_{s_c}$	$\{\ \star$	s_{g1}	$\star\ \}_{s_c}$
t_d	$\rightarrow$	$\{\ \star$	n_1	$n_3\ \}_{s_d}$	$\{\ \star$	t_g	$\star\ \}_{s_d}$	$\{\ \star$	s_{g2}	$\star\ \}_{s_d}$
...										

Triangle matrix after insertion

		0	1	2	3	4	5	6	7	8
...										
t_t	$\rightarrow$	n_p	n_2	n_0	t_n	t_b	t_g	2	s_b	0
t_g	$\rightarrow$	n_p	n_0	n_3	t_t	t_c	t_{n+1}	$\{\ 2$	s_c	0
...										
t_a	$\rightarrow$	$\{\ \star$	n_2	$n_1\ \}_{s_a}$	$\{\ \star$	t_n	$\star\ \}_{s_a}$	$\{\ \star$	1	$\star\ \}_{s_a}$
t_b	$\rightarrow$	$\{\ \star$	n_0	$n_2\ \}_{s_b}$	$\{\ \star$	t_t	$\star\ \}_{s_b}$	$\{\ \star$	1	$\star\ \}_{s_b}$
t_c	$\rightarrow$	$\{\ \star$	n_3	$n_0\ \}_{s_c}$	$\{\ \star$	t_g	$\star\ \}_{s_c}$	$\{\ \star$	1	$\star\ \}_{s_c}$
t_d	$\rightarrow$	$\{\ \star$	n_1	$n_3\ \}_{s_d}$	$\{\ \star$	t_{n+1}	$\star\ \}_{s_d}$	$\{\ \star$	1	$\star\ \}_{s_d}$
...										
t_n	$\rightarrow$	n_p	n_1	n_2	t_{n+1}	t_a	t_t	2	s_a	0
t_{n+1}	$\rightarrow$	n_p	n_3	n_1	t_g	t_d	t_n	2	s_d	0
...										

Table 6.3: Reconnecting a node on side 0 of a triangle.

t_d. Table 6.3 shows the element matrix before and after the splitting operation.

If the new node is found on side 1 or side 2 of the triangle similar reconnection patterns are to be used. These patterns are given in Appendix E.

The data movements have to be programmed in such a way that the information stored in the original triangle records is not overwritten if it is still required for completion of the records of the new triangles. It is recommended that for example a copy of the row of triangle t_t and its neighbour t_g in the element matrix is kept. The rows of the new triangles can then be filled with data from the copies.

6.6.3 Connecting nodes outside convex hull

If the new point which is to be inserted in the triangulation, lies outside the convex hull of the current triangulation a different reconnection strategy has to be adopted. The point will be connected to every node on the part of the convex hull which is visible from the new point. As discussed in Section 6.4.3 this is also the part of the hull which becomes obsolete when the hull is updated with the new point.

As shown on Figure 6.17 all new triangles will have the new point n_p as their starting node. The triangles will be given numbers t_n, t_{n+1}, ..., t_{n+f}. To find out the visible stretch of the hull to which the point will be connected the same algorithm is used as the one that updates the hull incrementally, which is given in Section 6.4.3.

In conjunction with Figure 6.17, Table 6.4 can be used to work out the adjacency information of the new triangles. It shows the rows in the element matrix that have to be added. Note that the first and the last of the new triangles have an adjacency pattern which is slightly different from the intermediate triangles.

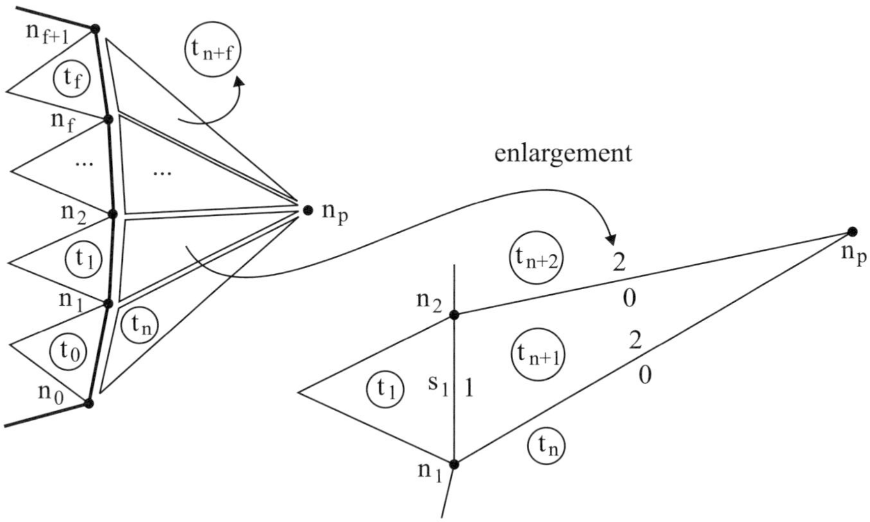

Figure 6.17: Reconnecting a node outside the convex hull.

For this procedure no copies of triangle records have to be made as no triangle numbers are overwritten. The only information which cannot be taken directly from the adjacency columns of the element matrix is the numbers of the triangles along the boundary which will have the new triangles as neighbours. To find out which triangle numbers these are, a small search has to be performed in the `tpnl` table. Once these numbers are known their adjacency over the boundary can be corrected to include the new triangles.

		0	1	2	3	4	5	6	7	8
...										
t_n	$\rightarrow$	n_p	n_1	n_0	t_{n+1}	t_0	-1	2	s_0	?
t_{n+1}	$\rightarrow$	n_p	n_2	n_1	t_{n+2}	t_1	t_n	2	s_1	0
t_{n+2}	$\rightarrow$	n_p	n_3	n_2	t_{n+3}	t_2	t_{n+1}	2	s_2	0
...										
t_{n+f}	$\rightarrow$	n_p	n_{f+1}	n_f	-1	t_f	t_{n+f-1}	?	s_f	0
...										

Table 6.4: Reconnecting a node outside the convex hull.

As the convex hull will have to be updated after insertion of this new point it is best to combine the procedures so the determination of the stretch of the obsolete convex hull is only determined once.

6.6.4 Edge flipping

Following the edge flipping strategy introduced by Lawson [16], all triangles, that are newly created have to be scrutinised and checked to see whether they violate the Delaunay criteria. A violating triangle triggers the start of an edge flipping series that will stop when the Delaunay integrity is recovered.

In the implementation presented here, a stack is used which contains all new triangles which have to be checked. The Lawson algorithm states that a newly inserted point will only have to be checked against the circumcircles of the neighbours to the newly created triangles. As a result of the layout of the reconnections patterns it is sufficient to store just the numbers of the newly created triangles since the new point is always the first node of the triangle and the relevant neighbour always lies on side 1 of the triangle.

The stack (called edge flip waiting list or `efwl`) is filled and read in a LIFO (last in first out) fashion. When a point is inserted all new triangles are added to the `efwl`. After the reconnection procedure is complete the edge flipping method commences. From the top of the stack a triangle is read. Node 0 of the triangle is checked against the circumcircle of the neighbouring triangle along side 1. If node 0 falls outside the circle the entry is removed from the `efwl` vector and the next entry is handled. Alternatively if node 0 falls inside the circumcircle the two neighbouring triangles are flipped and the resulting two triangles of the flipping procedure are both added to the `efwl` vector. The edge flipping stops when no entries are left in the `efwl` vector.

The actual flipping of two triangles must happen in such a way that the inserted point remains node 0 of the two resulting triangles after the flip. Figure 6.18 illustrates in symbolic terms how to organize this. Table 6.5 indicates the changes required in the element matrix.

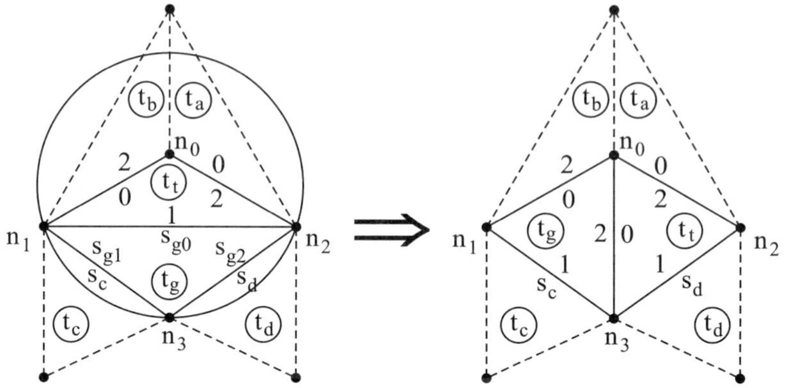

Figure 6.18: The edge flipping reconnection pattern.

In the example triangle n_0 is a newly inserted node which created three triangles t_t, t_a and t_b. Node n_0 is then found to be lying within the circumcircle of triangle t_g, the second neighbour (i.e. along side 1) of triangle t_t. The flipping procedure then decides to flip the diagonal in triangles t_t and t_g. As a consequence of the flipping not only the neighbourhood information for triangles t_a and t_b will have to be updated, but also that of triangle t_c and t_d, the two other neighbours of triangle t_g. The redefinition of triangles t_t and t_g is such that node n_0 becomes the first node of the new triangles. By doing this, the flipping procedure can be continued using the same reconnection pattern if for example node n_0 is then found to be lying within the circumcircles of either triangle t_c or t_d.

6.6.5 Creating a triangulation

To create a full triangulation of a series of nodes, be it boundary nodes or a set of initial nodes, all the steps explained above have to be combined. The incremental algorithm however has to be jump-started with the creation of an initial triangle. The first three nodes of the initial nodes, that are not colinear can be taken to form the initial triangle. As there are two ways that the three initial nodes can be connected so a clockwise test, as described in Section 6.4.3, is carried out to ensure that the initial triangle is oriented anti-clockwise.

Once the initial triangle is established all other nodes in the series are inserted one at the time using the incremental method.

Triangle matrix before edge flipping

	0	1	2	3	4	5	6	7	8
...									
$t_t \longrightarrow$	n_0	n_1	n_2	t_b	t_g	t_a	2	s_{g0}	0
$t_g \longrightarrow$	$\{ n_3$	n_2	$n_1 \}_{s_{g0}}$	$\{ t_d$	t_t	$t_c \}_{s_{g0}}$	$\{ s_d$	1	$s_c \}_{s_{g0}}$
...									
$t_a \longrightarrow$	n_0	n_2	$\star$	t_t	$\star$	t_b	2	$\star$	0
$t_b \longrightarrow$	n_0	$\star$	n_1	t_a	$\star$	t_t	2	$\star$	0
$t_c \longrightarrow$	$\{ \star$	n_3	$n_1 \}_{s_c}$	$\{ \star$	t_g	$\star \}_{s_c}$	$\{ \star$	s_{g1}	$\star \}_{s_c}$
$t_d \longrightarrow$	$\{ \star$	n_2	$n_3 \}_{s_d}$	$\{ \star$	t_g	$\star \}_{s_d}$	$\{ \star$	s_{g2}	$\star \}_{s_d}$
...									

Triangle matrix after edge flipping

	0	1	2	3	4	5	6	7	8
...									
$t_t \longrightarrow$	n_0	n_3	n_2	t_g	t_d	t_a	2	s_d	0
$t_g \longrightarrow$	n_0	n_1	n_3	t_b	t_c	t_t	2	s_c	0
...									
$t_a \longrightarrow$	n_0	n_2	$\star$	t_t	$\star$	t_b	2	$\star$	0
$t_b \longrightarrow$	n_0	$\star$	n_1	t_a	$\star$	t_g	2	$\star$	0
$t_c \longrightarrow$	$\{ \star$	n_3	$n_1 \}_{s_c}$	$\{ \star$	t_g	$\star \}_{s_c}$	$\{ \star$	1	$\star \}_{s_c}$
$t_d \longrightarrow$	$\{ \star$	n_2	$n_3 \}_{s_d}$	$\{ \star$	t_t	$\star \}_{s_d}$	$\{ \star$	1	$\star \}_{s_d}$
...									

Table 6.5: The edge flipping reconnection pattern.

6.7 Refinement strategies

The refinement strategy determines which procedure controlling the creation of new points in the triangulation until the mesh size criteria are reached. The point creation scheme is therefore the core of the mesh generation program. The two schemes which will be presented here are both iterative point insertion schemes. Refinement point after refinement point will be introduced into the triangulation until all triangles in the triangulation satisfy the criteria.

Both schemes are based on the automatic point creation procedure proposed by Weatherill *et al* [32]. Candidate refinement points are created automatically at the location of the centroids of existing triangles in the triangulation. The candidate points then have to pass two tests controlled by two parameters α and β. Only candidate nodes which pass both tests are accepted for insertion into the triangulation. The iterative procedure stops when there are no more candidates which pass both tests.

The first scheme is by Weatherill [32]. The second scheme is a variation by the authors that guarantees a stricter compliance with the mesh size requirements.

6.7.1 The first refinement scheme

The first refinement scheme is given in Algorithm 6.2. The algorithm starts from an initial triangulation of all boundary nodes.

Input: α, β – refinement parameters
Output: Modified triangulation to meet mesh size criteria
(1) `Refinement` := TRUE
(2) `refNods` := $\emptyset$; initialise refinement nodes
(3) **while** `Refinement` = TRUE
(4)   **foreach** triangle `t` inside the interior of the domain
(5)   `P` := the barycentre of `t`
(6)   `mpb` := determine the mesh parameter of point P
(7)   **foreach** $i \in \{0, 1, 2\}$
(8) $\mathtt{d}_i$:= distance between i-th vertex of `t` and point P
(9) $\mathtt{mpr}_i$:= mesh parameter of i-th vertex of `t`
(10)   **if** $\mathtt{d}_i < \alpha * \mathtt{mpr}_i$
(11)   **goto** Line 4; do not refine `t`
(12)   **foreach** node `n` $\in$ `refNods`
(13)   `d` := distance between node `n` and point P
(14)   **if** $\mathtt{d} < \beta * \mathtt{mpb}$
(15)   **goto** Line 4; do not refine `t`
(16)   `refNods` := `refNods` + $\{P\}$; accept point P
(17)   **if** `refNods` is empty
(18)   `Refinement` := FALSE; refinement is finished
(19)   **else**
(20)   **while** `refNods` is not empty
(21)   `P` := point from `refNods`
(22)   Add point `P` to the triangulation by the Delaunay Kernel

Algorithm 6.2: The first refinement scheme.

6.7.2 The second refinement scheme

The second refinement scheme is given in Algorithm 6.3. The algorithm starts from an initial triangulation of all boundary nodes.

Input: α, β – refinement parameters
Output: Modified triangulation to meet mesh size criteria

```
(1)   Refinement := TRUE
(2)   refNods := ∅; initialise refinement nodes
(3)   while Refinement = TRUE
(4)      foreach triangle t inside the interior of the domain
(5)         foreach side f_i (i = {0,1,2}) in triangle t
(6)            d := length of side f_i
(7)            if d > α ∗ min(mpr_j, j ∈ side points)
(8)               goto Line 4; do not refine t
(9)         P := the barycentre of t
(10)        mpb := determine the mesh parameter of point P
(11)        foreach node n ∈ refNods
(12)           d := distance between node n and point P
(13)           if d_j < β ∗ mpb
(14)              goto Line 4; do not refine t
(15)        refNods := refNods + {P}; accept point P
(16)     if refNods is empty
(17)        Refinement := FALSE; refinement is finished
(18)     else
(19)        while refNods is not empty
(20)           P := point from refNods
(21)           Add point P to the triangulation by the Delaunay Kernel
```

Algorithm 6.3: The second refinement scheme.

6.7.3 Discussion of the refinement schemes

The critical differences between the two refinement schemes presented here is how the mesh parameter is compared with the local length scale. This is called the α test.

When a candidate node in the first scheme is examined for insertion, a comparison is made between the distances from the candidate to the corner nodes and the local mesh parameters at the corner nodes. By default the candidate node is accepted until one of these connecting lines is shorter than the required mesh size. The philosophy behind this approach is to avoid creating too small edges in the triangulation. The mesh size therefore is interpreted as the radius of an imaginary circle around each node in the current triangulation in which no new points should be inserted.

The second scheme is not concerned about creating too small edges but tries to avoid all edges that are too large. In this scheme a candidate node passes the α test if it lies in a triangle which has at least one side that is longer than the mesh parameter of the candidate node. This scheme guarantees that there will be no edges left in the mesh that violate the mesh parameter values assigned to its two end nodes.

In the second test all candidate refinement nodes have to pass the β test. This test avoids refinement nodes which among themselves are too close to each other, being inserted in the same round of refinements.

Default values for α and β are both unity. Weatherill *et al* [31] came to the conclusion that β has an influence on the regularity of the final refined mesh, whereas α controls the tolerance on the mesh size criteria.

Alternative schemes (not implemented by the authors) could include:

- Examination of all edges in the current triangulation and insertion of a refinement node on all the edges which do not satisfy the mesh size criteria of their two end nodes [13].
- Insertion of a refinement node on the location of the centre of the circumcircle of each triangle whose side lengths exceed the mesh size criteria [6].
- Examination of all triangles which have at least one side too large for the mesh size criteria. Depending on how many sides are too large a refinement node is inserted alternatively at the centroid of the triangle, in the middle of the side which is too large or on an interpolated location between those two options (for example see the work of Muylle *et al* [22]). This approach also takes the status of the neighbouring triangles into account.

6.8 Post-processing tasks

A series of post-processing tasks have to be applied to finalize the generated mesh. This includes boundary recovery, domain recovery and finally smoothing.

6.8.1 Boundary recovery

When the triangulation of the boundary refinement nodes is completed it cannot be guaranteed for non-convex domains that all the segments that make the refined boundary are actually present in the triangulation as edges of triangles. To be able to mark all triangles as being either inside or outside the domain all boundary segments have to be recovered.

6.8.1.1 Recovering missing boundary segments

Boundary recovery consists of a checking procedure that verifies the presence of all the boundary segments in the triangulation. If certain segments are missing (because no triangles exist that have the segment as one of its edges), the segment can be brought back into the triangulation by either of the following procedures:

- Insertion of an extra boundary refinement node half way between the two nodes that define the missing segment. The two halves of the missing segment may now appear in the triangulation. But this is not guaranteed. Each missing half segment is bisected again in a recursive manner until the full length of the original boundary segment is present in the triangulation. This method recovers the boundary segments without compromising the Delaunay properties of the mesh. But a great disadvantage of the method is that it distorts the mesh size regularity by creating edges which are far smaller than necessary. This recovery procedure is often called *stitching*.

- Alternatively the topology of the mesh can be changed and missing segments retrieved by flipping the common edge between neighbouring triangles. As shown in Figure 6.19 the topology change depends on the configuration of the triangles involved in the missing segment problem. The most common occurrence of missing segments (case (a) in Figure 6.19) can be solved by a single edge flip. A more complicated (case (b)) in Figure 6.19) requires a few steps to resolve. This procedure does not distort the mesh regularity but it violates the Delaunay properties of the mesh.

Other procedures have been discussed by Weatherill [30].

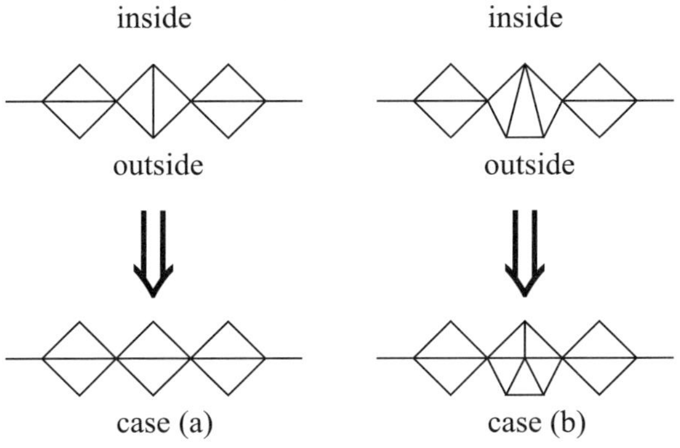

Figure 6.19: Two cases of boundary recovery.

In the implementation presented here only the trivial problem (case (a) in Figure 6.19) is handled by means of flipping. If the more complicated problem is detected stitching is used instead. Implementation wise, boundary recovery is not a very difficult procedure. A general loop goes through all segments of outer and inner boundaries. To check the presence of a segment the `tpnl` table is used. If there is at least one triangle that belongs to the `tpnl` entries of both the start and the end nodes of the segment the segment is present in the triangulation.

To use the flipping solution on a missing segment first the two triangles that are involved must be determined. The `tpnl` table is used here as well in conjunction with the adjacency information of the element matrix. A loop through all the triangles in the `tpnl` entry of the start node is undertaken to determine if they have a neighbour on the side opposite of the start node. If the neighbour has the end node as one of its nodes, the two triangles that will be involved in the flip have been detected. The edge flip is solved in the same way as the edge flipping procedure discussed in Section 6.6.4. Care has to be taken that all adjacency information and the `tpnl` records are correctly updated. If no two triangles can be found it is a sign that a more complicated missing segment situation has occurred.

Then the stitching procedure is used instead. Determination of the location of the halfway node and insertion of it into the triangulation is undertaken in the same way as the insertion of any other boundary refinement node. The boundary definition and convex hull vectors must be updated.

Figure 6.20 shows the flowchart associated with the boundary recovery procedure.

6.8.2 Removing out-of-domain elements

Once all segments of the boundary are present in the triangulation triangles can be marked as being *inside* or *outside* of the domain. If the domain is convex there are no outside triangles and all triangles in the triangulation can be marked as being inside. The inside-outside flag is stored in column 9 of the element matrix. If the domain is non-convex the following procedure is suggested to determine the value of the flag:

1. Initialise the flag for all triangles to value 0, which indicates that the triangle has not yet been considered and is currently undetermined.
2. For each segment of outer and inner boundary:
   - 2.1 Determine the triangle on the left and on the right hand side of the segment.
   - 2.2 Give the triangle on the left the flag value $+1$ and the triangle on the right (if there is one) -1.
3. While there are triangles remaining that have a value -1 and have undetermined neighbours (i.e. neighbours with a flag value 0):
   - 3.1 Set the value of the flag of these neighbours to -1.
4. Set the value of the flag of all remaining undetermined triangles to $+1$.

As there are usually far fewer outside triangles than inside triangles this procedure tracks down the outside triangles first, which are recognised easily as they share a segment of the boundary. It is therefore important that the outer boundary is oriented anti-clockwise and the hole boundaries clockwise so the interior of the domain always lies on the left hand side.

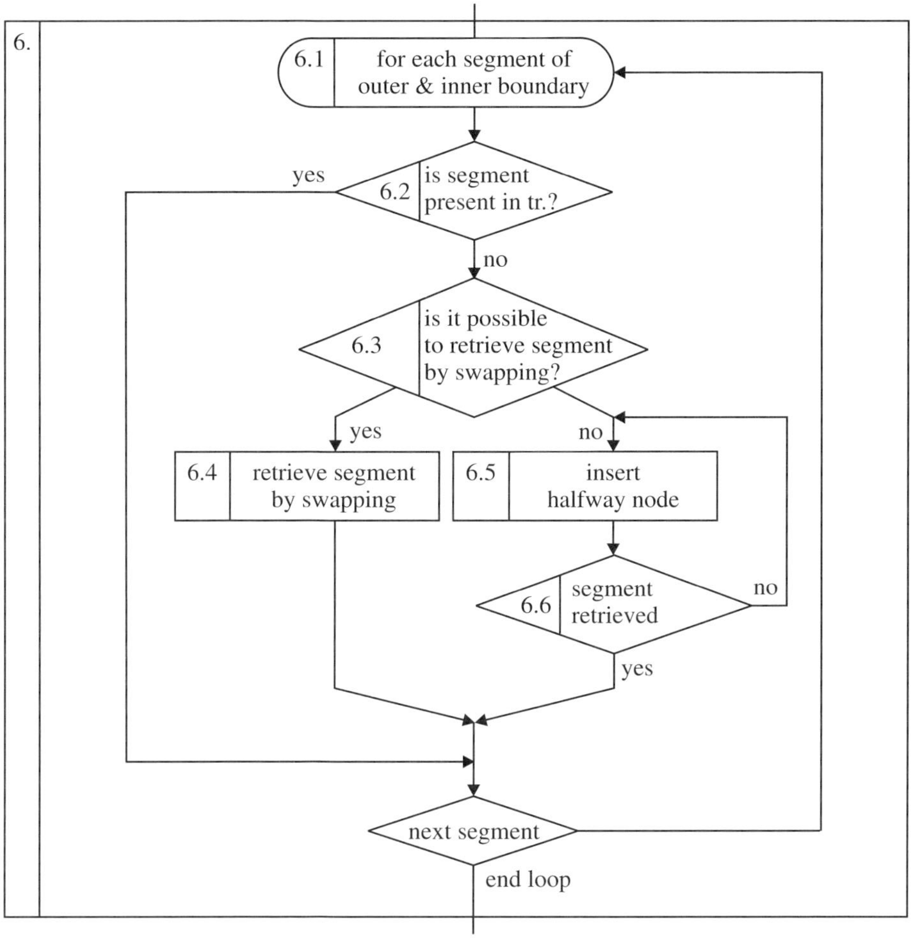

Figure 6.20: Flowchart for: 6. boundary recovery.

Once all outside triangles have been marked, all remaining undetermined triangles must be inner ones. It can be argued that the search for outside triangles with undetermined neighbours is quite an intensive procedure and that several loop operations have to be carried out. As shown in Figure 6.21 the damage however stays limited as a result of the fact that potentially only one layer of outside triangles can exist. Outside triangles are the result of meshing of the convex hull instead of only the non-convex domain. All the outside triangles therefore have only corner nodes that are part of the boundary. If they are not found in the first loop (because they are adjacent to the boundary), they must be found in the first round of checking undetermined neighbours.

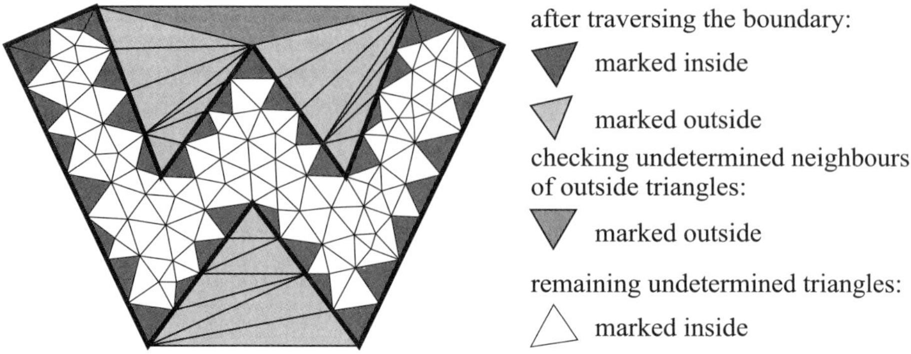

Figure 6.21: Detecting inside and outside triangles.

6.8.3 Smoothing

Smoothing is an operation in which the positions of all internal nodes are changed slightly in order to retrieve a mesh with better quality triangles. The connectivity of the mesh is not altered and neither are the boundary nodes. Changing the positions of the internal nodes might affect the Delaunay properties of the mesh. Therefore smoothing can only be applied when the generated mesh is final and does not have to be further refined. Chapter 9 is devoted to the subject of smoothing and mesh quality enhancement.

6.9 Concluding remarks

In this chapter the basics of the Delaunay triangulation method were introduced. Combined with a point creation scheme this method was then used to develop a mesh generation algorithm.

The following chapter will expand the Delaunay principles from two to three dimensions, where they will be applied to develop a tetrahedral mesh generation algorithm.

Bibliography

[1] Barber, C.B., "The Quickhull Algorithm for Convex Hulls", *ACM Transactions on Mathematical Software*, 22(4): 496–483, 1996.

[2] Borouchaki, H., George, P.L., "Aspects of 2D Delaunay mesh generation", *International Journal for Numerical Methods in Engineering*, 40: 1957–1975, 1997.

[3] Borouchaki, H., George, P.L., Lo, S.H., "Optimal Delaunay Point Insertion", *International Journal for Numerical Methods in Engineering*, 39: 3407–3437, 1996.

[4] Bowyer, A., "Computing Dirichlet tessellations", *The Computer Journal*, 24(2): 162–166, 1981.

[5] Chand, D.R., Kapur, S.S., "An algorithm for convex polytopes", *Journal of the ACM*, 17: 78–86, 1970.

[6] Chew, P.L., "Guaranteed-Quality Delaunay Triangulations", Technical Report TR-89-983, Dept. of Computer Science, Cornell University, 1989.

[7] Chew, P.L., "Guaranteed-Quality Mesh Generation for Curved Surfaces", in *Proceedings of the Ninth Annual Symposium on Computational Geometry (San Diego, California)*, pages 274–280. Association for Computing Machinery, May 1993.

[8] Day, A.M., Tracey, D., "Parallel implementation for determining the 2D convex hull", *Concurrency-Practice and Experience*, 10: 449–466, 1998.

[9] Delaunay, B., "Sur la sphère vide", *Bulletin of the Academy of Sciences of the USSR VII, Class Sci. Mat. Nat.*, 6: 793–800, 1934.

[10] Dirichlet, G.L., "Uber die Reduction der positiven quadratischen Formen mit drei understimmten ganzen Zahlen", *Z. Angew. Math. Mech.*, 40(3): 209–227, 1850.

[11] Formaggia, L., "Data Structures for Unstructured Mesh Generation", in Thompson, J.F., Soni, B.K., Weatherill, N.P., Editors, *Handbook of Grid Generation*, Chapter 14, pages 1–22. CRC Press, 1999.

[12] Fortune, S.J., "A sweepline algorithm for Voronoi diagrams", *Algorithmica*, 2: 153–174, 1987.

[13] George, P.L., Borouchaki, H., *Delaunay Triangulation and Meshing, Application to Finite Elements*, Editions Hermes, Paris, France, 1998.

[14] Green, P.J., Sibson, R., "Computing Dirichlet tesselations in the plane", *The Computer Journal*, 21(2): 168–173, 1978.

[15] Guibas, L., Stolfi, J., "Primitives for the Manipulation of General Subdivisions and the Computation of Voronoi Diagrams", *ACM Transactions on Graphics*, 4 (2): 74–123, 1985.

[16] Lawson, C.L., "Software for C1 Surface Interpolation", in Rice, J.R., Editor, *Mathematical Software III*, pages 161–194. Academic Press, New York, USA, 1977.

[17] Lee, D.T., Lin, A.K., "Generalized Delaunay triangulation for planar graphs", *Discrete & Computational Geometry*, 1(3): 201–217, 1986.

[18] Lee, D.T., Schachter, B.J., "Two algorithms for Constructing a Delaunay Triangulation", *Int. J. Comput. Inf. Sci.*, 9(9): 219–242, 1980.

[19] Marcum, D., Weatherill, N., "Unstructured Grid Generation Using Iterative Point Insertion and Local Reconnection", *AIAA Journal*, 33(9): 1619–1625, Sept. 1995.

[20] Marcum, D.L., "Unstructured Grid Generation Using Automatic Point Insertion and Local Reconnection", in Thompson, J.F., Soni, B.K., Weatherill, N.P., Edi-

tors, *Handbook of Grid Generation*, Chapter 18, pages 1–31. CRC Press, Boca Raton, USA, 1999.

[21] Mavriplis, D.J., "An Advancing Front Delaunay Triangulation Algorithm Designed for Robustness", paper 93-0671, AIAA, January 1993.

[22] Muylle, J., Iványi, P., Topping, B.H.V., "A new point creation scheme for uniform Delaunay triangulations", *Engineering Computations, International Journal for Computer Aided Engineering and Software*, 19(6): 707–735, 2002.

[23] O'Rourke, J., *Computational Geometry in C*, Cambridge University Press, 1993.

[24] Rupert, J., "A Delaunay Refinement Algorithm for Quality 2-Dimensional Mesh Generation", *Journal of Algorithms*, 18(3): 548–585, May 1995.

[25] Seed, G.M., "Delaunay and Voronoi tessellations and minimal simple cycles in triangular region and regular-3 undirected planar graphs", *Advances in Engineering Software*, 32(5): 339–351, 2001.

[26] Shamos, M.I., Hoey, D., "Closest-point problems", in *Proceedings of the 16th Annual Symposium on the foundations of Computer Science*, pages 151–162, October 1975.

[27] Shewchuk, J.R., *Delaunay Refinement Mesh Generation*, PhD thesis, Carnegie Mellon University, Pittsburgh, PA, USA, 1997.

[28] Voronoï, G., "Nouvelles applications des paramères continus à la théorie des formes quadratiques. recherches sur les parallélloèdres primitifs.", *Journal Reine angew. Math.*, 134, 1908.

[29] Watson, D.F., "Computing the n-dimensional Delaunay tesselation with application to Voronoi polytopes", *The Computer Journal*, 24(2): 167–172, 1981.

[30] Weatherill, N.P., "The reconstruction of boundary contours and surfaces in arbitrary unstructured triangular and tetrahedral grids", *Engineering Computations*, 13(8): 66–81, 1996.

[31] Weatherill, N.P., Hassan, O., Marcum, D.L., "Calculation of Steady Compressible Flowfields with Finite Element Method", Technical Report AIAA-93-0341, American Institute of Areonautics and Astronautics, 1993.

[32] Weatherill, N.P., Hassan, O., Marcum, D.L., "Compressible Flowfield Solutions with Unstructured Grids Generated by Delaunay Triangulation", *AIAA Journal*, 33(7): 1196–1204, 1995.

Chapter 7

Three-dimensional Delaunay triangulation

7.1 Introduction

In this chapter a three-dimensional Delaunay mesh generation is discussed. First some differences between the 2D and the 3D cases are discussed then an implementation strategy is explained. Next the general layout of the program used by the authors for this mesh generation is given. The basic definitions and notations used in this chapter are introduced in Section 7.3. Section 7.4 provides an outline of the Delaunay kernel and insertion polyhedron method. Section 7.5 is devoted to the design of data structures and it is followed by Section 7.9 that describes the use of those data structures within the Delaunay kernel implementation. Section 7.9.1 discusses an adaptive arithmetic package and robust geometric predicates used by the Delaunay kernel.

7.2 A comparison of the 2D and 3D approaches

The 3D Delaunay mesh generator program differs significantly from the 2D equivalent described in Chapter 6. The main differences are in: the approach to the input data; the algorithm of the Delaunay kernel; and the design of the data structures. These differences and some similarities are discussed in more detail in the next section.

7.2.1 Comparison of input data

The 2D mesh generator uses a coarse mesh as input data. The coarse mesh describes the geometry of a mesh domain and is used to construct boundaries of the domain. The boundaries are refined before triangulation of the domain interior takes place. In the 3D case the use of the coarse tetrahedral mesh would place a significant constraint on the geometries that could be handled and would make it difficult to prepare the input data. Instead of a coarse mesh the domain is described by the surface mesh of the domain boundaries. Additionally the input data can contain meshes of any 1D or 2D features embedded in the domain (e.g. internal boundaries). The input meshes should be the final ones, which means the boundaries are not refined or modified further. This approach has the advantage of making the design of the mesh generator less complex. On the other hand the drawback of this approach is that poor quality surface meshes may result in poor quality volume meshes. The alternative approach where the boundaries are meshed together with the interior of the domain can be found in reference [9].

7.2.2 Comparison of Delaunay kernels

In the 2D Delaunay triangulation described in Chapter 6 the edge–flipping algorithm has been used. Although it is possible to construct a 3D equivalent of that algorithm (see reference [6]), the polyhedron insertion method appears to be easier to implement. Some numerical problems arising in determination of the insertion polyhedron were solved by using the adaptive precision floating-point arithmetic package described in reference [8].

7.2.3 Comparison of data structures

The design of the data structures for the 3D program differs significantly from the one used in the 2D case described in Chapter 6. Generally speaking the 2D program utilizes three main arrays to store information for the input data and triangulation. These arrays are:

- the element array that stores information about element vertices, neighbours and element flags;
- the node array that stores information about elements adjacent to a node;
- the coordinate array that stores information about position of the nodes.

In the 3D program, instead of using arrays the information about the input meshes and a triangulation is distributed among collections of objects like tetrahedra, faces, edges, points. If care is taken to construct a proper interface to these objects their internal structure can be hidden and modified at any time without influencing the rest of the program.

Data hiding is not the only advantage of the "object oriented" design. The other important point is that the 2D triangulation process can be easily visualized and followed on screen from the beginning to end. This is a very valuable debugging tool. In the 3D case however even if visualization can be maintained it is hard to follow the process of triangulation. It may be easier to have data encapsulated as separate objects to check the correctness of the implementation of the object interface in a separate environment independently from the mesh generation program.

Another issue related to the above is that in most cases the descriptions of the algorithms are formulated in terms of elements, points, edges, faces, for example "... find the neighbouring tetrahedron to a given one ..." or "having an edge find all tetrahedra with this edge ...". In the authors' opinion the programming approach based on the encapsulation of data into objects makes it easier to translate the description of an algorithm into a program.

7.2.4 Comparison of refinement strategies

There are two stages of refinement in the current implementation.

- **Boundary surface refinement**: The mesh refinement strategy described in Chapter 6 can be easily generalized to the 3D case of surface refinement. This method is used in the current implementation where the boundary of the 3D mesh is defined by plane surfaces and the elements of the surface mesh are refined individually. The input data for the three-dimensional mesh generator is the refined surface mesh. The restriction of using solely plane elements exists only in the implementation presented here and the method itself does not rely on this fact. This means that curved, refined surfaces can also be used as input data to the mesh generator. The boundary surface refinement is not discussed further in this chapter.
- **Volume refinement**: Similarly to the 2D case, two different refinement criteria could be implemented. In the first case the element size distribution is driven by the mesh size parameters of the input boundary meshes. In the second case the element size is controlled by a background mesh. This latter approach has not been implemented here.

7.3 Definitions

This section introduces a number of definitions and the notation that will be used throughout this chapter. First some conventions about triangles and tetrahedra are presented, then definitions of triangulation, mesh and various element sets are described. The latter follows the work of George [4], where the reader can find more mathematical details.

7.3.1 Definition of triangles and tetrahedra

Triangles and tetrahedra are well known entities from elementary geometry. The description and geometric properties of those objects can be found in reference [13].

These will not be repeated here; however some assumptions about the numbering of vertices, edges and faces will be introduced. The special procedure for numbering those elements will help later on, for example in enumerating vertices of a face or finding end nodes of an edge shared by two faces.

The generic numbering of vertices and edges of a triangle is such that a vertex P_i lies opposite to the edge e_i ($i = \{0, 1, 2\}$), as shown in Figure 7.1. When viewing the triangle the vertices are numbered in an anti-clockwise manner. The order of the

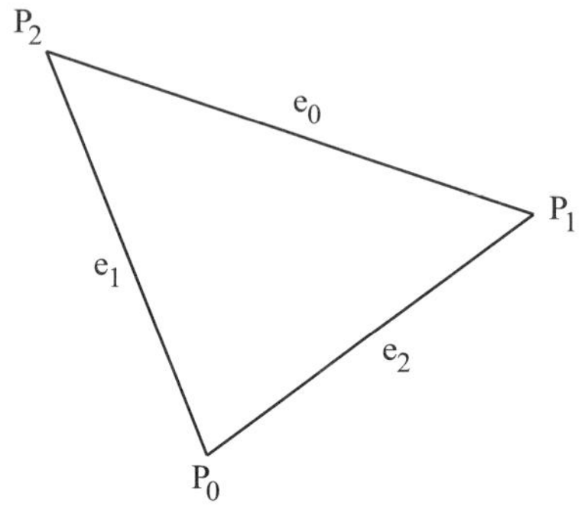

Figure 7.1: Generic numbering of vertices and edges in a triangle.

vertices defining the edges is assumed to be as follows:

$$\begin{aligned} e_0 &: P_1 \quad P_2 \\ e_1 &: P_2 \quad P_0 \\ e_2 &: P_0 \quad P_1 \end{aligned}$$

The generic numbering of vertices and faces of a tetrahedron is such that vertex P_i lies opposite to face f_i ($i = \{0, 1, 2, 3\}$), as shown in Figure 7.2. The order of the vertices in a face is assumed to be as follows:

$$\begin{aligned} f_0 &: P_1 \quad P_2 \quad P_3 \\ f_1 &: P_0 \quad P_2 \quad P_3 \\ f_2 &: P_0 \quad P_1 \quad P_3 \\ f_3 &: P_0 \quad P_1 \quad P_2 \end{aligned}$$

When looking at a face from outside the tetrahedron vertices appear in an anti-clockwise order.

7.3.2 Definition of triangulation and a mesh

The concept of a triangulation and a mesh appears repeatedly in this book. Sometimes those two terms are misused as synonyms. This section presents informal definitions of triangulation and mesh. It also points out the essential differences between them.

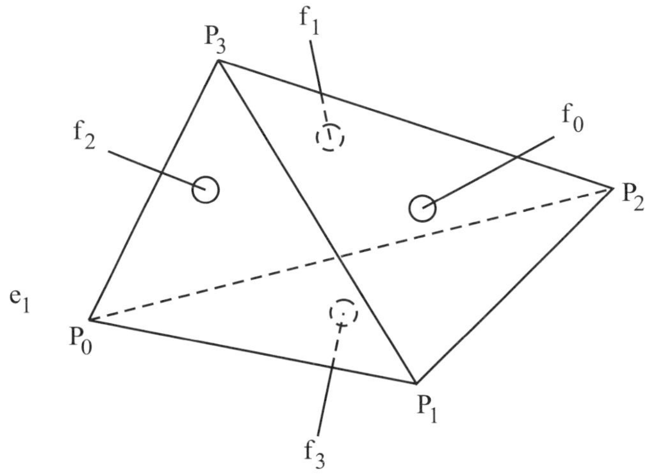

Figure 7.2: Generic numbering if vertices and faces in tetrahedron.

A triangulation is always defined with respect to a set of points (usually in R^2 or in R^3). A triangulation is such an arrangement of triangles (in 2D) or tetrahedra (in 3D) having vertices from the set, that the triangles or tetrahedra completely cover the convex hull of the set of points, with the restriction that the triangles or tetrahedra do not overlap and there are no gaps between them. The triangulation is said to be conforming if an intersection of any two triangles (tetrahedra) is a vertex, or a whole edge, or a whole face (in 3D) or an empty set. A mesh on the other hand is defined with respect to a closed bounded domain in 2D or 3D. A conforming triangulation of that domain is referred to as a mesh. The difference between these two concepts is illustrated in Figure 7.3.

7.3.3 Element sets

The mesh generation algorithm is described using various sets of elements. The definitions of these sets follow exactly the ones given in reference [4]. For the sake of clarity the concepts of the element sets will be illustrated in the 2D case. Practical algorithms of the construction of the element sets are presented in Section 7.9.

In the definitions below the following symbols are used:

$\mathcal{T}$ – conforming triangulation

P – a point distinct from all vertices in the triangulation

V – a vertex of the triangulation

K – an element in $\mathcal{T}$

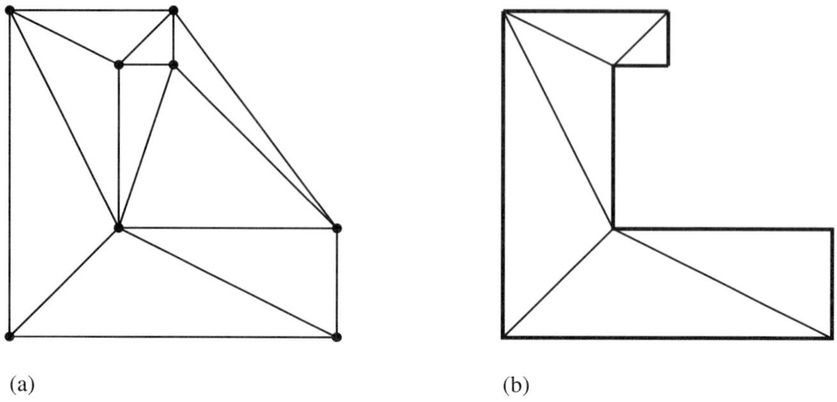

Figure 7.3: Illustration of triangulation and mesh concepts: (a) triangulation of a set of points, (b) mesh of the domain.

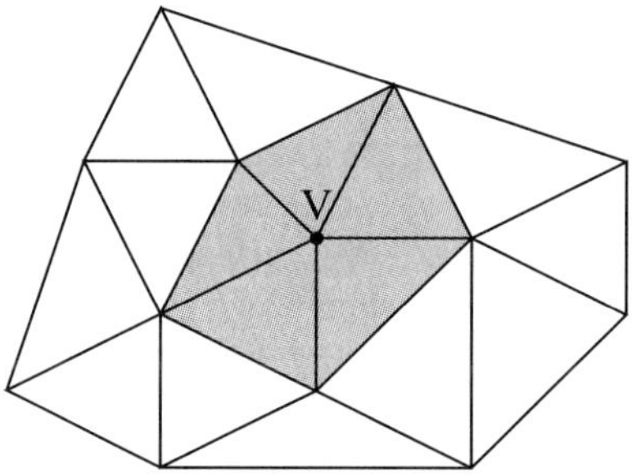

Figure 7.4: Ball $\mathcal{B}$ associated with vertex V.

- **Ball**: The ball $\mathcal{B}$ associated with vertex V is the set of elements which include V as a vertex. The ball is shown in Figure 7.4.

- **Base**: If point P (not a vertex) is enclosed in the triangulation, the base (B_P) associated with P is the set of elements in $\mathcal{T}$ enclosing P.
  The definition of base can be extended to include the case when P is not enclosed in a triangulation (see reference [4]). As a result of the fact that a reduced incremental method is used such extensions will not be necessary.

- **Circumdisc or circumball**: The circumdisc C_K associated with a given element K is the circle which can be defined by the three vertices of triangular element K. The circumball is the three-dimensional equivalent of the circumdisc, a sphere which is defined by the four vertices of a tetrahedra K.

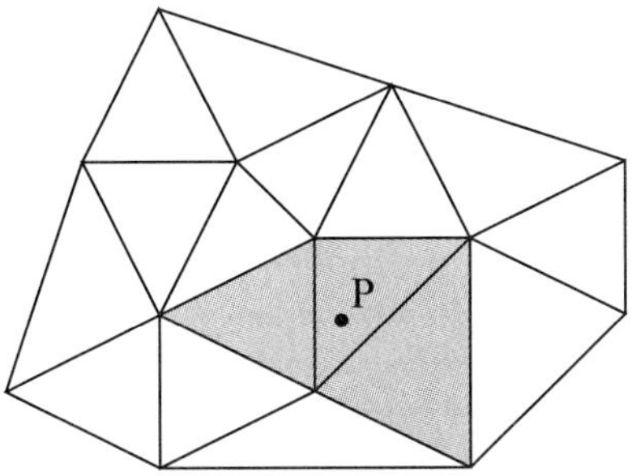

Figure 7.5: Cavity associated with point P.

- **Cavity**: The cavity associated with a given point P is the set of elements in $\mathcal{T}$ whose circumdiscs (circumballs) enclose the point P. A cavity is shown in Figure 7.5.
- **Pipe**: Let s be a segment enclosed in $\mathcal{T}$, assuming that the endpoints of s are two vertices in the triangulation. Then, in two dimensions the pipe associated with s is the set of elements having at least one edge intersected by s. A pipe is shown in Figure 7.6.

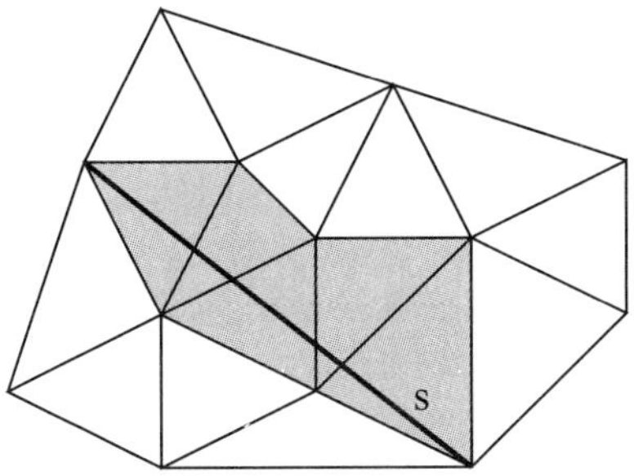

Figure 7.6: Pipe associated with segment s.

- **General pipe**: Let s be a segment enclosed in $\mathcal{T}$, assuming that the endpoints of s are two vertices in the triangulation. Then in three dimensions, the general pipe associated with s is the set of elements having at least one face or one edge intersected by s.
- **Edge shell**: Let e be an edge in $\mathcal{T}$. The shell associated with e is the set of elements sharing edge e.
- **Face shell**: Let f be a face in a three dimensional triangulation $\mathcal{T}$. The face shell associated with f is the set of one or two elements sharing that face.

7.3.4 Miscellaneous definitions

The following definitions are required for the description of the volume refinement method which follows the approach described by Li [7]:

- **Diametric sphere**: The diametric sphere of a boundary segment is the smallest sphere that encloses the segment.
- **Equatorial sphere**: The equatorial sphere of a triangle face is the smallest sphere that passes through the three vertices of the face.
- **Encroachment**: A boundary segment is encroached if its diametric sphere is not empty (for example contains a point of the triangulation). A boundary face is encroached if its equatorial sphere is not empty.
- **Picking region of a tetrahedron**: The picking region of a tetrahedron is a sphere which has the same centre as the circumsphere of the tetrahedron and its radius is proportionate to the radius of the circumsphere of the tetrahedron. The ratio between the radius of the picking zone and the radius of the circumsphere is $\delta < 1.0$.
- **Picking region of a triangle face**: The picking region of a triangle face is a circle which has the same centre as the circumcircle of the triangle and its radius is $\delta < 1.0$ times smaller than the radius of the circumcircle of the triangle face.
- **Picking region of a segment**: The picking region of a segment is a subsegment which has the same middle point as the segment and the length of subsegment is $\delta < 1.0$ times smaller than the length of the segment.

7.4 The Delaunay kernel and the reduced incremental method

As mentioned in Section 7.2.2 the mesh generator is based on the polyhedron insertion method of constructing a Delaunay triangulation. In particular it follows the algorithm first presented by Bowyer [3] and Watson [11].

This section presents the core procedure of the mesh generator based on the work of George [4] and referred to as the Delaunay kernel. The reader interested in a more comprehensive discussion of the Delaunay kernel as well as a strict mathematical proof of the method should consult references [2, 4].

Some remarks must be made at this stage about the description of the Delaunay kernel. First, according to the definition in Section 7.3.2, a triangulation $\mathcal{T}_i$ of the set of points P_k $k = 1, 2, \ldots, N$ is the convex hull of points P_k. Second, a triangulation is called a Delaunay triangulation if the circumdisc or circumball of all elements in triangulation $\mathcal{T}_i$ is empty. In other words the circumdisc or circumball of any of the elements in triangulation $\mathcal{T}_i$ does not contain any of the points P_k $k = 1, 2, \ldots, N$.

The purpose of the incremental Delaunay method is to include a new point P in the existing triangulation $\mathcal{T}_i$ in such a way that the new triangulation $\mathcal{T}_{i+1}$ is still a Delaunay triangulation. The procedure which constructs $\mathcal{T}_{i+1}$ from $\mathcal{T}_i$ and point P will be referred to as the Delaunay kernel.

However this method requires an initial Delaunay triangulation $\mathcal{T}_0$. The obvious choice for initial triangulation $\mathcal{T}_0$ is a single triangle in 2D or a single tetrahedron in 3D as these are very easy to construct. On the other hand more complex initial triangulations may improve the quality of the resulting mesh.

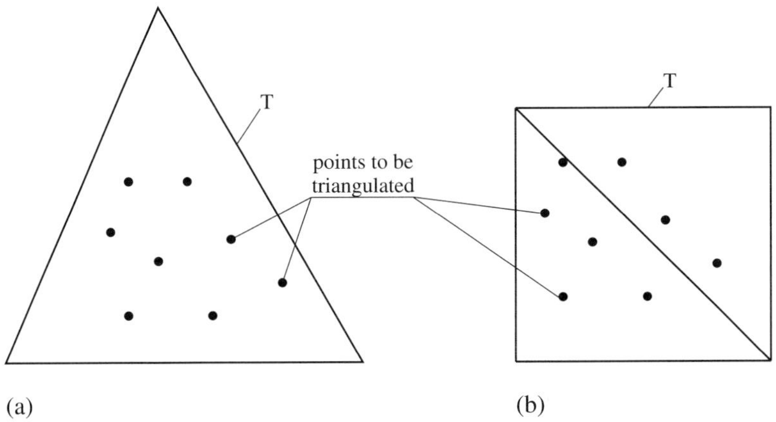

Figure 7.7: Examples of initial triangulation in 2D case.

When constructing new a triangulation $\mathcal{T}_{i+1}$ two cases should be considered: when point P is enclosed in $\mathcal{T}_i$; and when point P is outside $\mathcal{T}_i$.

A full discussion of both cases can be found in reference [4]. The authors have restricted their implementation to the reduced incremental method. This method ensures by proper selection of the initial triangulation $\mathcal{T}_0$, that each newly added point P will be enclosed within the current triangulation. This is possible if the initial triangulation $\mathcal{T}_0$ is determined in such a way that it completely encloses the convex hull of points being triangulated, see Figure 7.7. In practice this can be done by introducing a set of new points, for example the vertices of a triangle or tetrahedron or another easily triangulated shape which encloses the points being triangulated.

The basic algorithm of the Delaunay kernel can be expressed using the following equation [4]:

$$\mathcal{T}_{i+1} = \mathcal{T}_i - \mathcal{C}_P + \mathcal{B}_P \tag{7.1}$$

where:

$\mathcal{B}_P$ ball of point P;

$\mathcal{C}_P$ cavity associated with point P.

Thus in order to construct a new triangulation $\mathcal{T}_{i+1}$ which includes point P as a vertex the following steps are required (as also shown in Figure 7.8):

1. In triangulation $\mathcal{T}_i$ find the cavity associated with point P (as the implementation is restricted to the reduced incremental method, point P will always be enclosed within the current triangulation. Thus this cavity as defined on page 135 will never be empty).
2. Remove cavity elements from $\mathcal{T}_i$.
3. Connect point P with vertices forming external edges (faces in 3D) of the cavity.

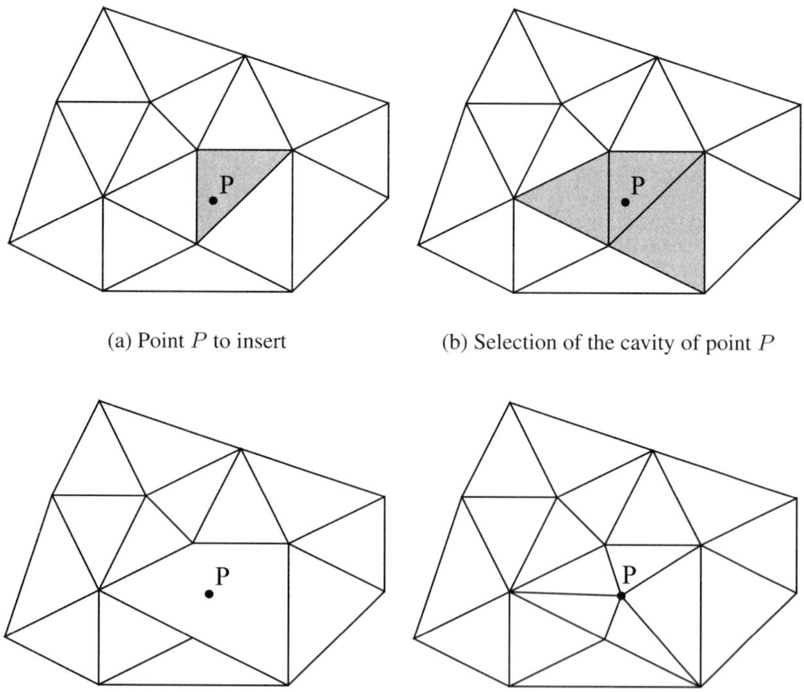

(a) Point P to insert

(b) Selection of the cavity of point P

(c) Removal of the cavity elements

(d) Reconnecting point P with the boundary of the cavity

Figure 7.8: Application of the Delaunay kernel.

7.5 Data structures

The specification of data structures and their precise implementation is essential as they influence the implementation of the Delaunay kernel and other procedures within the mesh generator. The specification is subject to several constraints including memory requirements, trade off between storing information and its recalculation and to some extent the language of implementation. It should also be noted that it is not sufficient to concentrate only on the Delaunay kernel as the data structures must be suitable for other procedures including the processing of input data, boundary recovery, mesh optimization. There must also be provision for data not related primarily to the mesh generation but to the other steps of the finite element analysis: material model data, decomposition data, etc.

The description of data structures is divided into two parts. First data structures related directly to the triangulation and the Delaunay Kernel are described followed by the presentation of data structures required by other parts of a mesh generator program.

7.5.1 Points and tetrahedra

Points and tetrahedra have been selected to be the basic building blocks of a triangulation. In general points store information about geometry (x, y, and z coordinates) while tetrahedral elements store information about mesh connectivity. In order to simplify the implementation of the algorithms some of the mesh connectivity information is also stored in the data structure of the mesh points. In particular the `Point` data structure contains:

- point coordinates x, y and z;
- nodal mesh parameter δ;
- vector of binary flags;
- list (stack) of elements having that mesh point as a vertex; and
- other information (meshpoint number, user flags, etc.).

The basic operations performed on the `Point` data structure include:

- creation of the `Point` objects;
- deletion of the `Point` objects;
- set/get operation retrieving coordinates, mesh parameter, flags; and
- adding/removing and searching for tetrahedra adjacent to the `Point` object.

The tetrahedral element data structure, `Tetrah`, contains

- four pointers to `Point` objects forming the element vertices;
- four pointers to adjacent tetrahedral elements (called neighbours); when there is no neighbouring tetrahedron at the given face it is indicated by special NULL pointer;
- a vector of binary flags; and
- other information (element number, user flags, element type, etc.).

The basic operation performed on the `Tetrah` data structure include:

- creation of the `Tetrah` object;
- deletion of the `Tetrah` object;
- set/get operation retrieving vertices, neighbours, flags, etc;
- checking if `Tetrah` object has a specified face;
- determine the edges shared by two given faces of the `Tetrah` object; and
- determine if two faces share a given edge of the `Tetrah` object.

Table 7.1 shows an example of the above data structures in the case of two dimensions. In the table instead of using tetrahedral elements triangular elements are used but the main concept of the data structure remains the same.

The list of elements in the `Point` structure provides a convenient way to enumerate all elements having the given point as a vertex. It would also be possible to carry out this enumeration using tetrahedron adjacency information, but the choice of either solution is a trade off between speed, memory usage and the ease of implementation.

The way the pointers to vertices and neighbouring elements are stored inside the `Tetrah` object depends on the implementation. The only requirement is that it should be possible to obtain the neighbouring element opposite to a given vertex and vice versa. In the implementation presented in this book the pointers to vertices and neighbours are stored in arrays. The special numbering of vertices and faces makes it easy to relate vertices to faces.

The next question is to decide in what data structures the `Points` and the `Tetrah` objects should be organized. The number of points and elements in a mesh is not known in advance. However the mesh generation process will require at least the following operations to be performed:

- the addition of new points to the triangulation (in the Delaunay kernel),

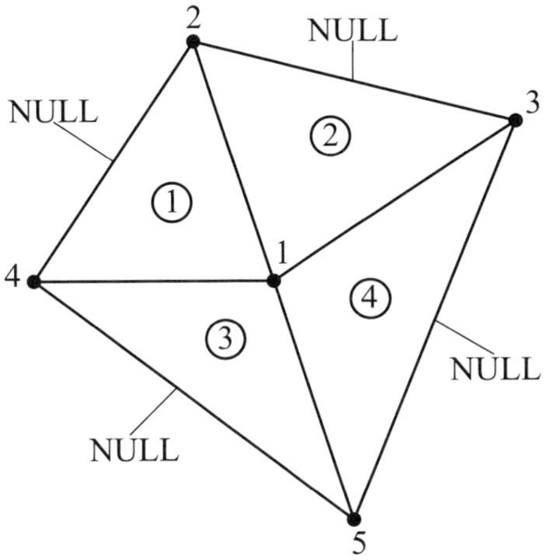

POINT 1
x_1, y_1 – coordinates
δ_1 – mesh parameter
$[1, 2, 3, 4]$ – adjacent elements

POINT 2
x_2, y_2 – coordinates
δ_1 – mesh parameter
$[1, 2]$ – adjacent elements

...

ELEMENT 1
$[1, 2, 4]$ – vertices
[NULL, 3, 2] – adjacent elements

ELEMENT 2
$[3, 2, 1]$ – vertices
[1, 4, NULL] – adjacent elements

...

Table 7.1: Basic data structures for the case of two dimensions.

- the removal of points from the triangulation (in the boundary recovery and mesh optimization),
- the addition of new elements to the triangulation (in the Delaunay kernel, boundary recovery and mesh optimization),
- the removal of elements from the triangulation (in the Delaunay kernel, boundary recovery and mesh optimization).

The above list indicates that it would be convenient to implement the collection of elements and the collection of points as a double linked lists. Linked lists facilitate easy insertion and removal of elements and they provide a simple and convenient way to handle dynamically changing data collections.

7.5.2 Tetrahedral mesh data structure

The next data structure to be introduced is the `TetrahMesh` structure which contains: the list of points; the list of tetrahedra; the number of points; and the number of tetrahedra. Although it may not seem to be necessary to introduce this data structure, however from the point of view of the program design it is convenient to encapsulate the lists of points and tetrahedra in one object. In addition to the two lists a `TetrahMesh` object contains two counters – the counter for the points in the mesh and the counter for the tetrahedra elements.

7.5.3 Edges and Faces

The `Edge` and `Face` data structures are provided to facilitate the handling of the data input, triangulation and boundary recovery. The `Edge` and `Face` objects are used not only as a convenient way of passing collections of two or three vertices but they also permit the assignment of additional information to the objects they represent. The `Edge` data structure consists of:

1. two pointers to the edge endpoints;
2. two pointers to faces sharing that edge;
3. a vector of binary flags;
4. other finite element information; and
5. a list of pointers to edges generated on the original boundary during the boundary recovery process.

Points 2 and 5 need some clarification. First of all the pointers to faces sharing the edge are used only in the context of input surface meshes. In a triangulation an edge can be shared by more than two faces, however it is assumed that in the input meshes an edge is shared at most by two faces! The last member in the structure is also necessary, since during boundary recovery some additional points may be inserted on the edge in order to recover the particular edge. The insertion of the point on the boundary edge splits the edge and the original edge keeps a record of all of the sub-edges. However if a sub-edge is further split by a point then the created new edges are stored with the sub-edges and not with the original boundary edge. Initially the list of sub-edges for all edges is empty indicating that the edges are not split. The `Face` data structure consists of:

1. three pointers to the face vertices;
2. three pointers to the face edges;
3. a vector of binary flags;
4. other finite element related information; and
5. a list of pointers to faces which fully cover the original face.

The `Face` objects are created primarily as a result of the parsing of input surface meshes. During boundary recovery if a face is missing from the triangulation it is recovered as a sum of sub-faces, which themselves are `Face` objects. The last member in the structure keeps track of all subfaces generated during the face recovery phase. The list of subfaces is initially empty.

7.5.4 The mesh generator data structure

The input to the mesh generator consists of series of surface meshes. Face and Edge data structures are used to hold the information for the input meshes. For each input mesh the following collection of data is maintained: a table of points in the mesh; a table of faces in the mesh; and a table of edges in the mesh. These data structures are encapsulated in one data structure called `MeshGenData`. In the program there exists only one instance of that data structure at a given time. The `MeshGenData` structure also contains some other mesh generation related information such as: mesh generation control information, the generated mesh, etc.

7.6 Program layout

In this section a general layout of the program is presented. As a result of the differences between data structures and Delaunay kernel for the 2D and the 3D cases the layout of the 3D mesh generator differs from the mesh generator for the 2D case. The top-level flowchart of the program is shown in the Figure 7.9.

∘ 1. **load & parse job description**
This step is exactly the same as in the 2D case. The program loads a file which contains all instructions that indicate what kind of meshing operation has to be carried out and what the mesh input files are.

∘ 2. **load surface meshes**
In this step a series of input surface meshes are read from the specified files. The implementation of this step depends on the format of the mesh data file. In the authors' current implementation the `mdf` mesh data file format (see Appendix D.1) is used with each mesh being kept in a separate file.

∘ 3. **parse surface meshes into internal data structures**
In step 2 input meshes are read into a data structure which facilitates finite element programming. However meshes have to be transferred into the data structures of simple geometric objects (points, edges, faces) which directly support the mesh generation algorithm. For example element-to-element or node-to-element adjacency information stored implicitly in input meshes is extracted and saved explicitly in data structures for the mesh generator. Section 7.7 will present further details concerning the input data.

∘ 4. **construction and triangulation of a cuboid**
The reduced incremental method (see reference [4]) implemented in the program requires an initial triangulation which encloses the domain to be triangulated. The initial triangulation has been selected to be a five element triangulation of a cuboid enclosing the mesh domain. The construction and triangulation of a cuboid is discussed in Section 7.8.

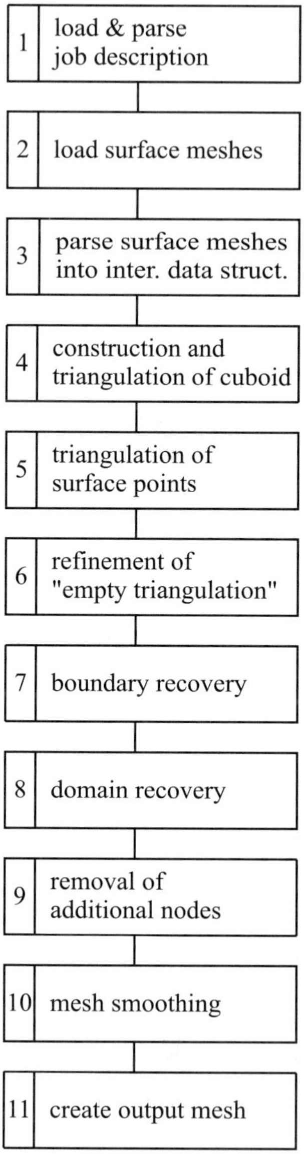

Figure 7.9: Top-level flowchart of 3D mesh generator.

- ∘ 5. **triangulation of surface points**
  All surface points are added one by one to the initial cuboid triangulation by the Delaunay kernel. The resulting triangulation is a so called "**empty triangulation**". Some more details are provided in Section 7.9.7.

- ∘ 6. **refinement of the "empty triangulation"**
  The "empty triangulation" is refined until the size criteria imposed by the mesh parameters are met. The new mesh points are added to the triangulation by the Delaunay kernel as described in Section 7.9 and 7.9.8.

- ∘ 7. **boundary recovery**
  The refined mesh is usually not boundary conforming – that means not all edges and surfaces from the input meshes are present in the triangulation. The refined mesh is modified to include all missing elements. The boundary recovery procedure is discussed in Section 7.10.

  - ∘ **7.1** **recovery of boundary edges**
    In order to recover the boundary surfaces of the mesh domain the edges of the boundary triangles have to be recovered first which is described in more details in Section 7.10.3.

  - ∘ **7.2** **recovery of boundary faces**
    The presence of all edges from boundary surfaces in the triangulation does not guarantee that all surfaces are present. Missing surfaces are recovered by the algorithm described in Section 7.10.4.

- ∘ 8. **domain recovery**
  The refined mesh domain is enclosed by the cuboid. Once the domain boundaries have been recovered it is possible to remove all elements which are outside the mesh domain. A procedure to recover the meshed domain from the cuboid is described in Section 7.11.

- ∘ 9. **removal of additional nodes**
  During the recovery of the boundaries of the mesh domain additional points may be added. These points can split edges or faces of the input surfaces. In this step these points are removed from the mesh. (It must be noted that this operation may not be possible in all cases.)

- ∘ 10. **mesh smoothing**
  The resulting mesh is smoothed according to the smoothing method selected in the mesh generation job description. The applicable smoothing techniques are described in Chapter 9.

- ∘ 11. **create output mesh**
  The final mesh is transformed from the internal data structures of the program into a mesh data structure indicated in the mesh generation job description.

7.7 Input data

The input to the mesh generator consists of a list of meshes composed of triangular and segment elements. It is required that:

- the first mesh is a triangular mesh that describes a closed surface that completely encloses other meshes and that
- meshes cannot touch or intersect each other as this will result in a non-conforming tetrahedral mesh.

Other input data to the mesh generator are the refinement method and the refinement parameters α and β. For the detailed input data file see Section C.2 in Appendix C.

The input meshes and input data are parsed into the `MeshGenData` structure. For each input mesh the following data structures are created: a table of meshpoints; a table of faces; and a table of edges.

7.7.1 Parsing boundary points

The boundary points of the input meshes are stored in a table of `Point` objects. Each `Point` object stores the coordinates of the point and the corresponding mesh parameter. In the objects a `POINT_BOUNDARY` flag is also set to help to distinguish between meshpoints coming from the input meshes and meshpoints added to the triangulation during the refinement phase. The flags `POINT_INT_BOUNDARY` and `POINT_EXT_BOUNDARY` help to distinguish between points coming from the external mesh or the internal meshes. The above flags proved to be useful during the boundary recovery phase and during removal of unwanted elements outside the meshing domain.

7.7.2 Parsing boundary faces

Boundary faces are handled in a similar way to boundary points. For each triangular face in each input mesh a `Face` object is created. Vertices of the `Face` object are set to the appropriate `Point` objects. The `Face` objects are stored in a table for each input mesh. This table is used in the boundary recovery phase of the mesh generation. The face–edge adjacency information is created during the parsing of the boundary edges.

7.7.3 Parsing boundary edges

During this phase of the processing of input meshes all edges in the meshes are enumerated and saved in a table of `Edge` objects. The edges are the boundaries of triangular elements. To enumerate all edges only once an adjacency "matrix" is created. When all of the edges of an element are considered it is marked by the FACE_BOUNDARY flag. Any element which encounters an adjacent element with the FACE_BOUNDARY flag does not store the edge shared between the current element and the marked, adjacent element again. At this phase the face–edge and edge–face adjacency information is also established.

7.7.4 General remark on input data processing

Algorithms for parsing input meshes into the native data structures of the mesh generator depend on the format and information carried by the input meshes. The input meshes are in the mdf data file format of the E-Library (see Appendix D.1). In this file format the number of edges is not specified (as in many other mesh file formats).

7.8 Construction and triangulation of the cuboid

The reduced incremental method (see reference [4]) implemented in the program requires an initial triangulation which encloses the domain to be triangulated. The initial triangulation has been chosen to be a five-element triangulation of a cuboid enclosing the mesh domain.

7.8.1 Construction of the cuboid

It is a basic assumption in the program that all input meshes must be enclosed in the first input mesh therefore it is enough to scan only the table of the meshpoints of the first mesh ($\mathtt{Tp}_0$) for extreme coordinates (x_{low}, y_{low}, z_{low} and x_{high}, y_{high}, z_{high} of the meshpoints). Once the extreme coordinates are known eight Point objects can be created to form the vertices of a cuboid. The coordinates of these points are equal to the extreme coordinates of the mesh domain scaled by a value. The scaling is necessary to ensure that all points from the input meshes fall inside the cuboid. The scaling factor in the current implementation is two. Figure 7.10 shows the numbering of the cuboid vertices. The flag POINT_CUBOID is also set in the cuboid vertices to help to distinguish between these vertices and vertices of the input meshes and the additional vertices generated during the mesh refinement. This flag is also used during the mesh refinement phase to avoid refinement of tetrahedra that are outside the meshing domain.

7.8.1.1 Triangulation of the cuboid

After the cuboid vertices have been created and added to the list of the meshpoints in the TetrahMesh object the cuboid is triangulated. Figure 7.11 shows the trian-

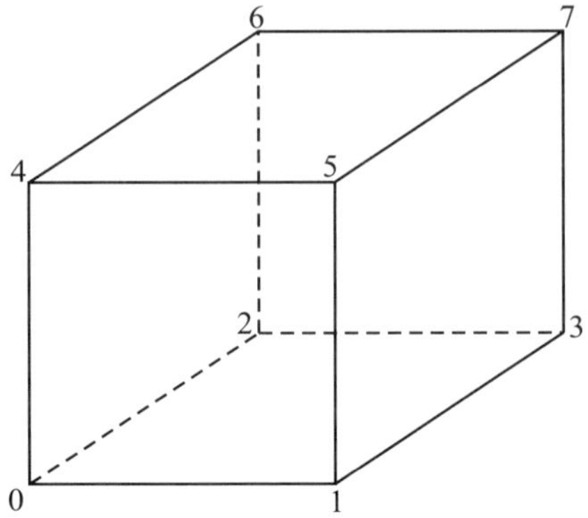

Figure 7.10: The numbering of the cuboid vertices.

gulation of the cuboid with five tetrahedra. The adjacency information of the created

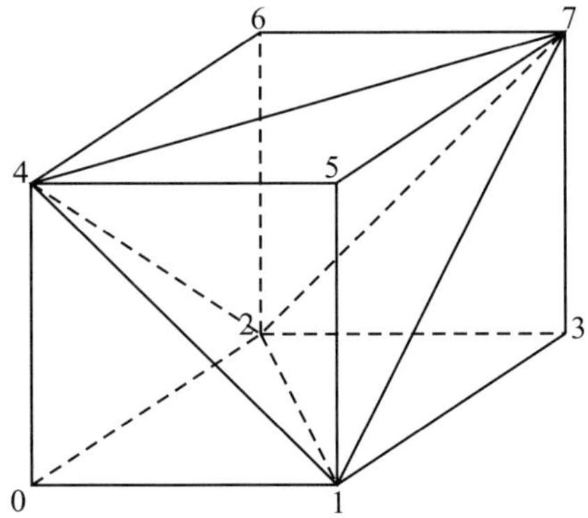

Figure 7.11: The triangulation of the cuboid.

tetrahedra is also established at this phase. To facilitate the triangulation and the setup of the adjacency relations a special matrix is introduced:

$$A = \left[\begin{array}{cccc|cccc} 0 & 1 & 2 & 4 & 4 & -1 & -1 & -1 \\ 4 & 5 & 7 & 1 & -1 & 4 & -1 & -1 \\ 1 & 3 & 2 & 7 & -1 & 4 & -1 & -1 \\ 7 & 6 & 4 & 2 & -1 & 4 & -1 & -1 \\ 7 & 4 & 2 & 1 & 0 & 2 & 1 & 3 \end{array}\right] \tag{7.2}$$

The matrix is based on the basic assumption of the numbering of vertices and faces of tetrahedron introduced in Section 7.3.1 and the numbering of the cuboid vertices. The first four columns describe the vertices of a tetrahedron while the next four columns describe the adjacency relation. The value '-1' in the matrix has been used to indicate that the face of the tetrahedron is not shared with any other tetrahedron, it is therefore an external boundary face.

After the triangulation of the cuboid the `TetrahMesh` object holding the generated mesh contains eight meshpoints and five tetrahedra. Adjacency relations are set for all tetrahedra and each meshpoint also holds a list of adjacent elements. This step provides a mesh which is a valid Delaunay mesh of the cuboid.

7.9 Implementation of the Delaunay procedures

This section presents the implementation of all of the Delaunay procedures: the Delaunay kernel, the triangulation of initial mesh points and the volume refinement of the mesh. The section also discusses some of the auxiliary procedures used by the Delaunay kernel:

- a procedure which examines the position of a point with respect to a tetrahedron;
- a procedure for finding the base of a point;
- a procedure for enumerating edge shell; and
- a procedure for constructing the cavity associated with a given point.

7.9.1 Robust geometric predicates

The Delaunay kernel procedure which inserts new points into the triangulation commonly performs a geometric test to determine whether a given point is enclosed in the circumball of a given tetrahedron or not. One way to express the test is to calculate the Delaunay measure [4]. It is assumed that P is the point under consideration, T stands for a tetrahedron, O_T is the circumcentre of T and r_T is the circumradii of T. The distance between point P and circumcenter O_T is denoted by $d(O_T, P)$. In this case the Delaunay measure of point P with respect to tetrahedron T is the ratio

$$\alpha(P,T) = \frac{d(P, O_T)}{r_T} \tag{7.3}$$

If $\alpha(P,T) < 1$ that means the point P is enclosed within the circumball of tetrahedron T. Unfortunately the geometric test based on the Delaunay measure may fail if the quantities involved in the Delaunay measure are not precisely calculated. This will result in an invalid cavity construction (see reference [4] for examples). In reference [4] several solutions to the ambiguity of this geometric test have been considered.

In the implementation developed here it was decided to use an approach in which circumcentres and circumradii are not calculated or updated. To determine if a point belongs to the circumball of a tetrahedron the determinant $\Delta(P,T)$ is calculated:

$$\Delta(P,T) = \begin{vmatrix} P_x & P_y & P_z & P_x^2 + P_y^2 + P_z^2 & 1 \\ v1_x & v1_y & v1_z & v1_x^2 + v1_y^2 + v1_z^2 & 1 \\ v2_x & v2_y & v2_z & v2_x^2 + v2_y^2 + v2_z^2 & 1 \\ v3_x & v3_y & v3_z & v3_x^2 + v3_y^2 + v3_z^2 & 1 \\ v4_x & v4_y & v4_z & v4_x^2 + v4_y^2 + v4_z^2 & 1 \end{vmatrix} \tag{7.4}$$

where $v1, v2, v3, v4$ are the vertices of the tetrahedron T. If $\Delta(P,T) < 0$ then point P belongs to the circumball of the tetrahedron T. This test is also subjected to numerical errors unless the coordinates are of integer-type and the calculation is exact. However instead of using integer-type coordinates the authors decided to use the Adaptive Precision Floating-Point Arithmetic and Fast Geometric Predicates package by J. Shewchuk. This package is a public domain and at the time of writing could be obtained from `http://www.cs.cmu.edu/~quake/robust.html`.

The package offers fast software-level algorithms for exact addition and multiplication of arbitrary floating-point values. Using these algorithms, fast robust adaptive geometric predicates have been built. The adaptivity of these predicates means that their running time depends on the degree of uncertainty of the result but is usually small. The package offers 2D and 3D orientation and in-circle tests. The orientation tests return positive, negative or zero values depending on the position of the point with respect to an oriented line in 2D (plane in 3D) which passes through specified points. The incircle tests return positive, negative or zero values depending on the position of the point with respect to the oriented circle in 2D (sphere in 3D) which passes through specified points. The 3D orientation test is used to determine the position of a point with respect to a tetrahedron (WHEREINTETRAH procedure) while the incircle test is used in the cavity construction algorithm (FINDCAVITY procedure).

7.9.2 Position of point in a tetrahedron

The position of a point P in relation to a tetrahedron T falls in one of the five categories:

- the point P lies inside the tetrahedron;
- the point P lies on a face of the tetrahedron;
- the point P lies on an edge of the tetrahedron;
- the point P is one of the vertices of the tetrahedron; or
- the point P lies outside the tetrahedron.

To determine the category to which a point P belongs; the signed volumes of four pairs of tetrahedra must be compared. In a pair of tetrahedra the first tetrahedron is made up by point P and one of the faces of tetrahedron T. The second tetrahedron in a pair is constructed from the same face as considered in the first tetrahedron in the pair and the remaining vertex in tetrahedron T. In the case of the second tetrahedron, though it is equivalent to tetrahedron T, it is important that the orientation of the topology is the same as for the first tetrahedron in the pair, since signed volumes are compared. The algorithm can be summarised as follows:

Input: Tetrahedron T; Point P;
Output: Position of Point P, index `fi` of face which either separates point P from vertices of T or on which point P lies, `sfi` sum of indices of faces on which point P lies.

WHEREINTETRAH(P, T)

`sfi` $:= 0$
`counter` $:= 0$
foreach tetrahedron face, `fi` $:= i$, $(i = \{0, 1, 2, 3\})$
    find vertices v_1^i, v_2^i, v_3^i of face `fi`
    find vertex v_4^i opposite to face `fi`
    $V_1 :=$ signed volume of tetrahedron $T_1 = \{v_1^i, v_2^i, v_3^i, P\}$
    $V_2 :=$ signed volume of tetrahedron $T_2 = \{v_1^i, v_2^i, v_3^i, v_4^i\}$
    if $V_1 = 0$
        `counter` $:=$ `counter` $+ 1$
        `sfi` $:=$ `sfi` $+ i$
        if `counter` $= 3$
            break
    else if $V_1 * V_2 < 0$
        `counter` $:= -1$
        break
switch `counter`
case -1
    return (P lies outside, `fi`, `sfi`)
case 0
    return (P lies inside, `fi`, `sfi`)
case 1
    return (P lies on face, `fi`, `sfi`)
case 2
    return (P lies on edge, `fi`, `sfi`)
case 3
    return (P lies on vertex, `fi`, `sfi`)

Algorithm 7.1: Position of point with respect to a tetrahedron.

In the case of point P falling outside tetrahedron T, `fi` is the index of the face which separates point P from the vertices of the tetrahedron. In the case of point P falling either on the face, edge or vertex, `fi` is an index of one of the faces containing point P and `sfi` is the sum of the indices of all faces containing point P. The face, edge or vertex on which point P falls can be determined by recalling the convention of the numbering of the faces and vertices and using the variables `fi` and `sfi`.

For example, assuming that point P falls on the edge shared by face three (vertices $\{0,1,2\}$) and face two (vertices $\{0,1,3\}$). Then

$$\texttt{fi} = 3$$
$$\texttt{sfi} = 5$$

To find the end nodes n_1 and n_2 of the shared edge, one can use the following scheme:

$$\begin{aligned} f_1 &= \texttt{fi} = 3, \\ f_2 &= \texttt{sfi} - \texttt{fi} = 5 - 3 = 2, \\ n_1 &= \min\{i : i \in \{0,1,2,3\} \text{ and } i \notin \{f_1, f_2\}\} = 0, \\ n_2 &= 6 - f_1 - f_2 - n_1 = 6 - 3 - 2 - 0 = 1. \end{aligned}$$

If point P falls on a vertex and

$$\texttt{fi} = 3$$
$$\texttt{sfi} = 4$$

the vertex number is

$$n = 6 - \texttt{sfi} = 2.$$

7.9.3 Search for the base associated with a given point

As defined in Section 7.3 base B_P is a set of elements enclosing the given point P. Since the implemented method is the reduced incremental method, point P is always enclosed in the current triangulation. Thus the base can consist of one element if P falls inside a tetrahedron, two elements if P falls on a tetrahedron face and of an arbitrary number of tetrahedra if point P falls on an edge. The case when P falls on a vertex will be regarded as an error, since this would mean point duplication in the mesh.

The three-dimensional version of the walking algorithm described in Section 6.6.1 is used to find the base of point P. The algorithm is implemented in the FINDBASE function. It should be emphasized again that as it is discussed in reference [4] the algorithm fails to make progress because of the occurrence of a cyclic path. The remedy for this is to modify the algorithm in such a way that the order in which the faces of the tetrahedra are checked changes randomly in each call to the procedure.

7.9.4 Enumeration of an edge shell

The enumeration of an edge shell is relatively simple using the lists of elements adjacent to mesh vertices (see the description of `Point` data structure in Section 7.5).

Given the edge end points A and B it is enough to visit all tetrahedra having, for example, point A as a vertex and check if they also have point B as a vertex. Those tetrahedra having both A and B form the edge shell.

7.9.5 Construction of the cavity associated with a given point

The construction of the cavity associated with a given point P is based on the fact that all elements of the base B_P belong to the cavity of P. Thus the cavity can be initialized by the base elements and then, by the use of element to element adjacency, the neighbouring elements can be visited. Each visited element is marked and a check is performed to determine if point P belongs to its circumball. This check is done by one of the geometric predicates described in Section 7.9.1. If P is enclosed in the circumball of the element the element is added to the cavity. The search ends when it is not possible to find any more elements not marked as visited using the adjacency relationship.

Input: Point P, the base B_P of point P
Output: `cavity(P)`, the list of elements whose circumball encloses point P and all out of cavity elements are also marked

FINDCAVITY()

(1) `cs` := Push all tetrahedra from B_P on checking stack `cs`
(2) **while** checking stack `cs` is not empty
(3)   `t` := pop element from checking stack `cs`
(4)   mark tetrahedron `t` as visited
(5)   **if** point P in circumball of `t`
(6)   `cavity(P)` = `cavity(P)` + {`t`}
(7)   **foreach** neighbour $\mathtt{tn}_i$ of `t` ($i = \{0, 1, 2, 3\}$)
(8)   **if** $\mathtt{tn}_i$ **not** marked as visited
(9)   push $\mathtt{tn}_i$ on checking stack `cs`
(10)   **else**
(11)   mark `t` as an element outside the cavity

Algorithm 7.2: Cavity of point P .

As the result of the algorithm, the list of elements which belong to the cavity of point P are made available and all elements adjacent to the external faces of the cavity are marked. The marking of elements outside the cavity helps to locate them when point P is connected to an external face of the cavity. Care must be taken when the cavity is created. In special circumstances point P may be very close to the circumball of an element and a decision must be made on whether the element belongs to the cavity or not. The result of an incorrect decision is the creation of an invalid cavity producing an invalid mesh. George [4] suggested several solutions to this problem. In the implementation presented here, any point which may lead to an ambiguous decision is not inserted into the mesh and moreover the algorithm relies on the precise geometric predicates to avoid the creation of invalid cavities.

7.9.6 The Delaunay kernel

As mentioned in Section 7.4 the Delaunay kernel is a procedure which allows the construction of a new triangulation $\mathcal{T}_{i+1}$ from the existing triangulation $\mathcal{T}_i$. The new triangulation $\mathcal{T}_{i+1}$ includes the new point P as a vertex while preserving the Delaunay property (empty circumballs of all tetrahedra). By using the algorithms described earlier the implementation of the Delaunay kernel is:

```
Input: Point P, current triangulation T_i
Output: Triangulation T_{i+1} with point P included as a vertex
DELAUNAYKERNEL(P, T_i)
    s := starting tetrahedron for the search of base of point P
    b := FINDBASE(P, s)
    cavity(P) := FINDCAVITY(P, b) (see algorithm 7.2)
    foreach element c_i (i = 1,...) in cavity cavity(P)
        foreach face f_j (j = {0,1,2,3}) of c_i
            Find neighbour n_j adjacent to face f_j
            if n_j = NULL or n_j is flagged as out of cavity
                Create new tetrahedron w
                Set vertices of w to vertices of face f_j and point P
                Push w on the stack of new elements newcav
    foreach element c_i (i = 1,...) in cavity cavity(P)
        foreach face f_j (j = {0,1,2,3}) of c_i
            Find neighbour n_j adjacent to face f_j
            Remove element c_i from the adjacency structure of n_j
            op := vertex opposite of face f_j
            Remove element c_i from the adjacency list of vertex op
        Remove element c_i from the mesh
    foreach element w_i (i = 1,...) in new cavity newcav
        foreach face f_j (j = {0,1,2,3}) of w_i
            ot := tetrahedron with face f_j
            if if ot exist
                Update adjacency of ot and w_i pointing to each other
        foreach vertex v_j (j = {0,1,2,3}) of w_i
            Add element w_i to the adjacency list of vertex v_j
        Add element w_i to the mesh
```

Algorithm 7.3: The Delaunay kernel .

The selection of the tetrahedron used at the start of the base search depends on the context of the Delaunay kernel.

- When the so called "empty triangulation" [4] is constructed the starting tetrahedron is selected as the most recently added element to the mesh.

- When the "empty triangulation" is refined the starting tetrahedron is selected as the element which is to be refined.

It must be noted that the operation of finding a tetrahedron with a given face (Line 20 in the algorithm) can be easily implemented with the vertex–tetrahedron adjacency lists in the `Point` structures. Each vertex of the given face has a list of tetrahedral elements and basically the procedure is looking for the intersection of these three lists. The result can only be empty, in the case of a boundary, or one tetrahedron.

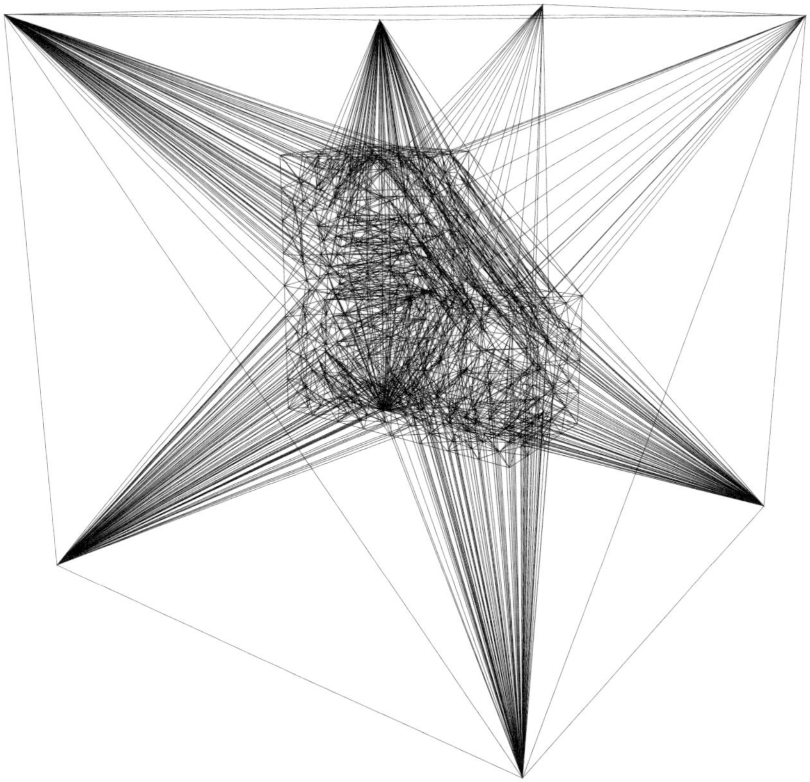

Figure 7.12: The cuboid and the "empty triangulation".

7.9.7 Triangulation of the input mesh points

After the cuboid mesh has been created the next step is to add all points from the input meshes to the initial triangulation. The points are added to the triangulation in the order they are stored in the input meshes by the Delaunay kernel. It has been shown that inserting the boundary points of the input mesh in a random order can improve both the efficiency and the mesh quality. At the same time it is relatively easy to implement the suppression of a point insertion if the insertion would result in ambiguous cavity creation (for example because of cospherical arrangements of points) and by trying to add the point later, hoping that inserting other points will resolve the problem. In the current program the randomised insertion is not implemented. However, the delayed insertion of "problematic" points is implemented. The result of this step is the so called "empty triangulation". Figure 7.12 shows the triangulation after all boundary vertices were added.

7.9.8 Volume mesh refinement

The mesh refinement in the 3D case follows the same philosophy as the mesh refinement in the 2D case described as Algorithm 6.3 in Section 6.7.2 with one exception, the position of the inserted points. In the case of 2D Delaunay refinement the algorithm inserts a new point at the centroid of the triangle. However in the 3D case the position of the inserted point is determined by Algorithm 7.5 as presented by Li [7]. This algorithm relies on Algorithm 7.4 to select a point for insertion which will produce "good" quality tetrahedra in the triangulation. However Algorithm 7.5 may modify the position of the selected point if it is too close to the boundary surface.

Input: object `obj`, number of points n, size of picking zone δ
Output: Point P
GOODPOINTSELECTION(n, δ)
(1) **foreach** $i = 1, \ldots, n$
(2) Randomly select point c_i in the picking zone of object `obj`.
(3) Find the cavity of point c_i.
(4) **foreach** tetrahedron w_j ($j = 1, \ldots, k$) which would be created by inserting point c_i
(5) Calculate the quality ω_j of tetrahedron w_j.
(6) Calculate $\rho_i = \min_{j=1,\ldots,k}(\omega_j)$, which is called *insertion quality.*
(7) Select Point $P = c_i$ which maximises the insertion quality ($\max_{i=1,\ldots,n}(\rho_i)$).
(8) **return** (Point P)

Algorithm 7.4: Selection of a "good" point.

```
Input: tetrahedron T
Output: Point to insert
POINTSELECTION(T)
    point A = GOODPOINTSELECTION(T, n, δ) (see Algorithm 7.4)
    if if point A encroaches on a boundary segment s
        point B = GOODPOINTSELECTION(s, n, δ) (see Algorithm 7.4)
        return (point B)
    if if point A encroaches on a boundary face f
        point C = GOODPOINTSELECTION(f, n, δ) (see Algorithm 7.4)
        if if point C encroaches on a boundary segment s
            point D = GOODPOINTSELECTION(s, n, δ) (see Algorithm 7.4)
            return (point D)
        return (point C)
    return (point A)
```

Algorithm 7.5: Selection of a point to be inserted into the triangulation.

Refinement points are created and inserted into triangulation until all tetrahedra meet the required mesh size criteria. As in the 2D case, the refinement process is governed by two parameters α and β. For the calculation of the mesh parameter at a candidate point P, Equation 5.13 is used.

7.10 Boundary recovery

The procedure for boundary recovery follows the approach proposed in reference [12]. After the triangulation has been finished not all edges and faces specified in the input meshes are present in the triangulation. If an edge or a face is not present in the triangulation it means that it intersects some of the tetrahedra in the triangulation. The recovery procedure searches for those intersection points and inserts them directly into the mesh modifying any tetrahedra related to the intersection point. In reference [4] this process is referred to as a method of constraint partitioning because the process recovers the missing edges or faces in the triangulation as a sum of sub-edges or sub-faces, respectively. In the constraint partitioning method, all edges need to be recovered first. When all boundary edges are present in the triangulation boundary faces can be recovered. The step of recovery of boundary faces is necessary as the presence of boundary edges in the triangulation does not guarantee that all faces will be present. For example in Figure 7.13 edges $\overline{v_1, v_2}$, $\overline{v_2, v_3}$ and $\overline{v_3, v_1}$ are present but face Δ_{v_1,v_2,v_3} is not. After all boundary edges and faces are recovered it is possible to remove additional intersection points by performing local remeshing of the balls related to those points.

Generally the local mesh modifications involved in edge and face recovery can be

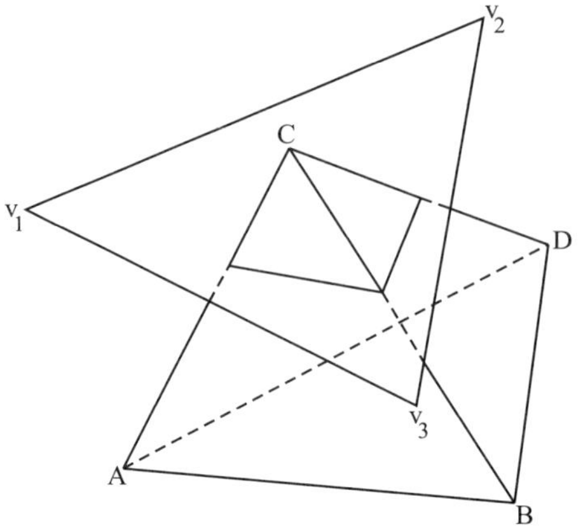

Figure 7.13: Edges are recovered but the face still is not present in the triangulation.

reduced to two procedures:

- adding a point on the edge thus splitting it into two sub-edges; and
- adding a point on the face thus splitting it into three sub-faces.

As these procedures do not use the Delaunay kernel, the triangulation after boundary recovery does not necessarily satisfy the Delaunay criteria.

7.10.1 Adding a point to a face

The procedure of adding a point to a face is relatively simple as it involves at most two tetrahedra adjacent to that face. If it is assumed that point P is inserted on face $f = \{v1, v2, v3\}$ of a tetrahedron $T = \{v1, v2, v3, v4\}$ then tetrahedron T is removed and three new tetrahedra are created (see Figure 7.14):

$$T \Longrightarrow \begin{cases} N_1 = \{v1, v2, v4, P\} \\ N_2 = \{v1, v2, v3, P\} \\ N_3 = \{v1, v3, v4, P\} \end{cases}$$

The tetrahedron T_a adjacent to T through face f is processed in a similar way. The process ends with the update of the node-to-element adjacency information in the vertices of T and T_a and with an update of the element-to-element adjacency information in the new tetrahedra and their neighbours.

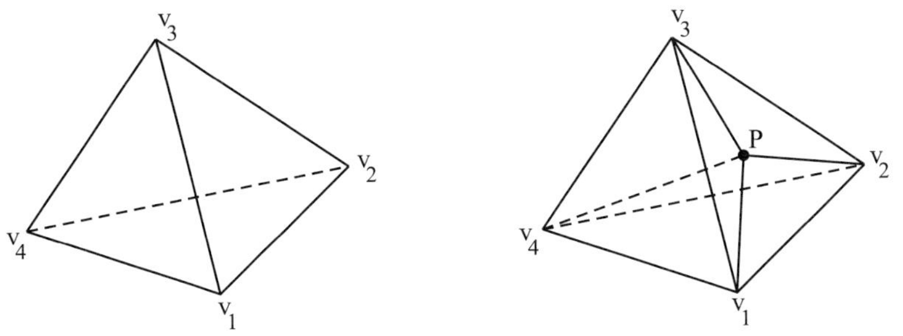

Figure 7.14: Adding a point to a face.

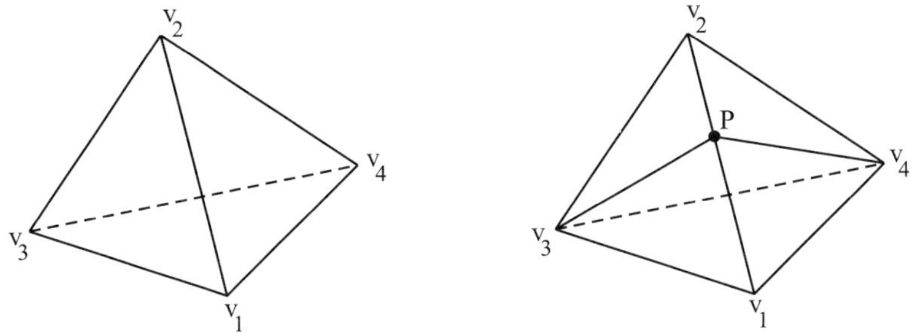

Figure 7.15: Adding a point on an edge.

7.10.2 Adding a point to an edge

The procedure of adding a point to an edge is more complicated than adding a point to a face as it can involve any number of tetrahedra (an edge shell can consist of one or more tetrahedra). If it is assumed that point P is inserted on edge $e = \{v1, v2\}$ of a tetrahedron $T = \{v1, v2, v3, v4\}$ then tetrahedron T is removed and two new tetrahedra are created (see Figure 7.15):

$$T \Longrightarrow \begin{cases} N_1 = \{v1, v2, v3, P\} \\ N_2 = \{v1, v3, v4, P\} \end{cases}$$

The same process is repeated with any tetrahedron which belongs to the shell of edge e. The process ends with the update of the node-to-element adjacency information in the vertices of all tetrahedra in the edge shell of e and the element-to-element adjacency information in the new tetrahedra and their neighbours.

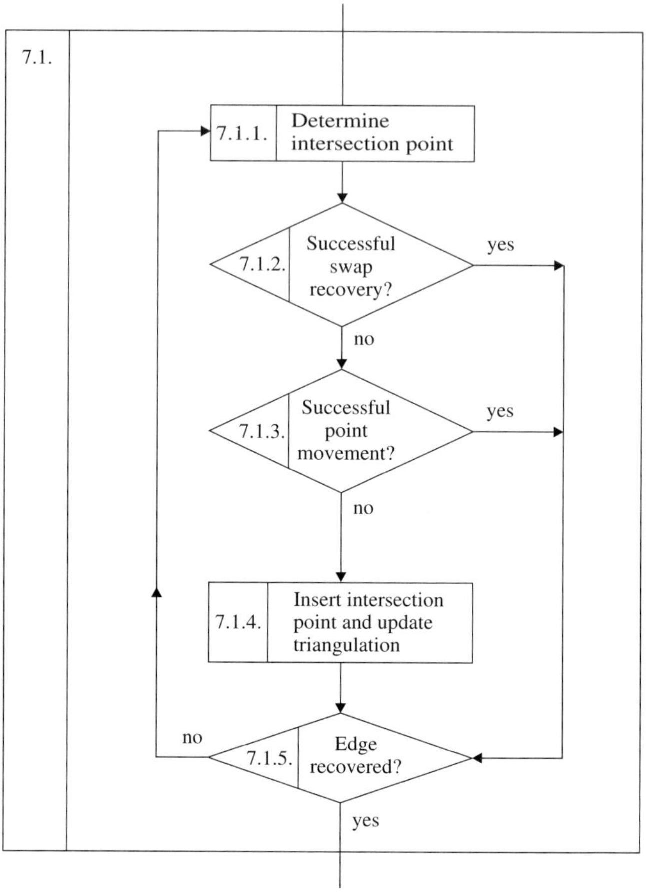

Figure 7.16: Recovery of missing boundary edges.

7.10.3 Recovery of boundary edges

The first step is to determine which edges of the boundary meshes are missing from the triangulation. The search for the edges missing from the triangulation is quite simple. An edge is present in the triangulation if its two end points are vertices of at least one tetrahedron. Thus to check if an edge is present in the triangulation it is enough to examine a ball of one of its end points and check if the second end point is the vertex of at least one tetrahedron from the ball. If no tetrahedron element can be found with the edge, it means that the edge is missing from the triangulation. If an edge is missing from the triangulation it is saved on a stack of missing edges. In the next step for each edge on the stack of the missing edges, the edge recovery algorithm is performed. The process of recovery of missing boundary edges is outlined in Figure 7.16.

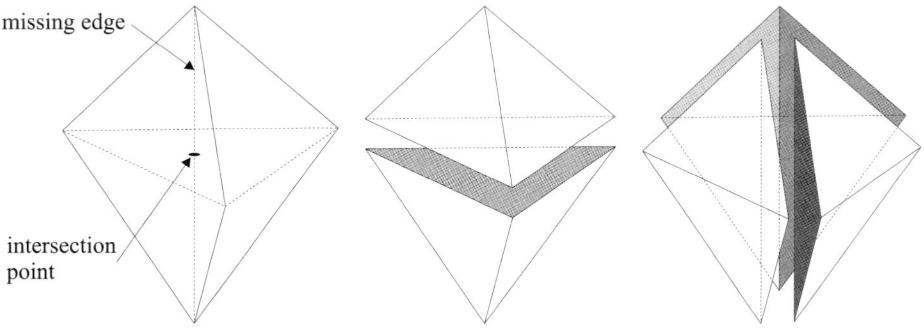

(a) Initial configuration (b) Initial configuration, exploded view (c) Swapped configuration, exploded view

Figure 7.17: Procedure for edge recovery by swapping edges.

If an edge is missing from the triangulation it means that it intersects either faces, edges or vertices of some tetrahedra. Depending on the intersection point there are several options to recover the edge:

- **Step 7.1.2. Swap recovery**
  When the edge intersects the face of a tetrahedron the edge may be recovered by swapping as shown in Figure 7.17. Before using this recovery technique, a check that the total volume of the elements before and after the swapping is the same must be performed.

- **Step 7.1.3. Point movement**
  It is possible to recover the missing edge by moving a point into a position which is on the missing edge. However this edge recovery can only be performed if the total volume of the tetrahedra elements before and after the recovery are the same and if the length criteria for the refinement is not violated. Figure 7.18 shows an example of this configuration in two dimensions. By moving point C the edge between points A and B can be recovered by the segments of $\overline{AC}$ and $\overline{CB}$.

- **Step 7.1.4. Insertion of intersection point**
  When there is no other way to recover the edge a new point can be inserted onto the edge. In this case the algorithm starts from one of the endpoints of the edge, for example A, and an intersection point P closest to that endpoint is determined. The procedure for the determination of the intersection points between an edge and a tetrahedron is discussed in Section 7.10.3.1. Then point P is added to the triangulation, which means that the edge between points A and P is now present in the triangulation.

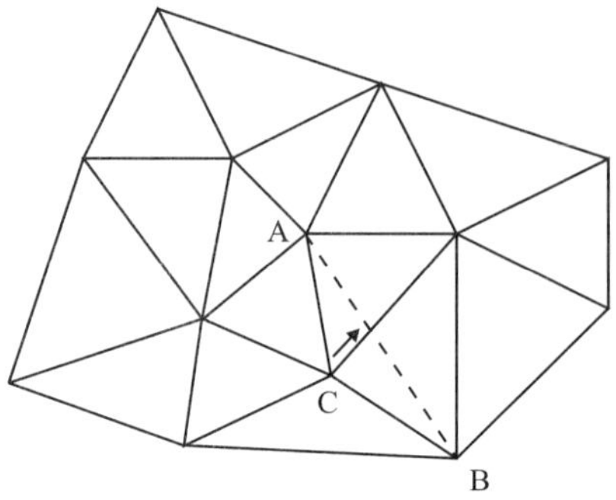

(a) Edge AB is missing

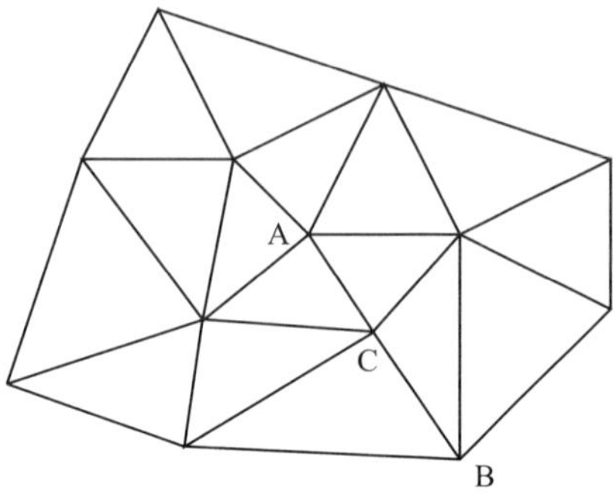

(b) Edge AB is recovered by the segments of $\overline{AC}$ and $\overline{CB}$

Figure 7.18: Procedure for edge recovery by moving a point.

- **Step 7.1.5. Edge recovered ?**
  At this point the whole edge may be recovered, but in certain cases further intersection points must be determined. In this case the procedure is repeated to recover the edge between point P and point A.

7.10.3.1 Intersection between segment and tetrahedral mesh

The main component of the edge recovery algorithm is the procedure which determines the intersection between a segment and the components of a tetrahedral mesh (faces, edges, vertices). It is assumed that the endpoints of the segment are vertices of the triangulation and the procedure determines only one intersection point at a time which is closest to the starting endpoint of the segment. As the segment is enclosed in the triangulation such an intersection point always exists.

If it is assumed that the endpoints of the segment are vertices $v1$ and $v2$, then in order to find the intersection point which is closest to vertex $v1$ it is enough to consider only the tetrahedra which form the ball of $v1$. Moreover it is enough to examine only those faces of the elements of the ball which do not include $v1$ as their vertex. The outline of the procedure is as follows:

1. find the ball of vertex $v1$
2. for every element in the ball
   (a) select a face f which does not contain vertex $v1$
   (b) check if there is an intersection between segment $v1$–$v2$ and face f.

There are several algorithms for segment-triangle intersection determination [1,5,10]. In the program developed by the authors the approach described in reference [1] is implemented. The intersection point is determined in two steps:

- First the equations for the segment line and triangle plane are formulated. Let V_0, V_1, V_2 be the vertices of the triangle A, B the endpoints of the segment. The equation of the triangle plane and the segment can be written in the form:

$$\overrightarrow{N} \cdot \overrightarrow{r} + d = 0 \qquad \text{plane equation} \tag{7.5}$$

$$\overrightarrow{r}(t) = \overrightarrow{A} + \overrightarrow{D}t \qquad \text{line equation} \tag{7.6}$$

  where:

$$\overrightarrow{N} = \overrightarrow{V_0V_1} \times \overrightarrow{V_0V_2} \qquad \text{normal to the triangle plane} \tag{7.7}$$

$$d = -\overrightarrow{V_0} \cdot \overrightarrow{N} \tag{7.8}$$

$$\overrightarrow{D} = \overrightarrow{B} - \overrightarrow{A} \qquad \text{direction vector of the line} \tag{7.9}$$

  The parameter t corresponding to line-plane intersection point is determined from the equation:

$$t = -\frac{d + \overrightarrow{N} \cdot \overrightarrow{A}}{\overrightarrow{N} \cdot \overrightarrow{D}} \tag{7.10}$$

  A check can be performed if the intersection point (if it exists) is on the segment ($t \in [0, 1]$). Having the t parameter the coordinates of the intersection point can be easily calculated from the line equation.

- In the second step the triangle and intersection point are projected on a coordinate plane x-y, x-z or y-z whichever is more orthogonal to the vector normal of the triangle plane. Then the position of the intersection point $\vec{r} = (x, y)$ is expressed in terms of two vectors formed by sides of the projected triangle $\vec{u}$ and $\vec{v}$:

$$\vec{r} = \alpha\vec{u} + \beta\vec{v} \tag{7.11}$$

  From the above equation the parameters α and β can be calculated and they can be used to determine the position of the intersection point with respect to the triangular face.

The intersection point can fall either on a face, an edge or a vertex. The procedure returns the intersection point coordinates and information on which category of intersections (face, edge, vertex) the intersection point belongs.

7.10.4 Recovery of boundary faces

The main algorithm for the face recovery differs from the algorithm for edge recovery. In the case of face recovery once a missing face is found it is recovered immediately.

The check if face f is present in the current triangulation is performed in a similar manner to a check for missing edges. Taking vertex v_i of face f all tetrahedra belonging to the ball of vertex v_i are examined to check if the other two vertices of face f are also the vertices of any of the tetrahedron from the ball. Face f is missing from the current triangulation if no tetrahedron element is found which includes all three vertices of face f. The algorithm of the face recovery is a modified version of the approach presented in reference [12].

To describe the algorithm, it is first assumed that face f is missing from the triangulation. The algorithm first collects all points on the edges of face f into set s. Set s will consist of the face vertices and any points added onto the edges of face f during the edge recovery. As the edges keep a record of any point added to them the collection of these points is straightforward. Once set s is initialised the algorithm considers each point P in set s. The ball of point P, $\mathcal{B}_P$, is examined to check if edges of elements T of the ball intersect face f. This check is carried out by looking at how many vertices are shared between tetrahedron T from the ball and set s. The following cases are possible:

1. tetrahedron T has one or two vertices which are in set s, which indicates that some edges of tetrahedron T may intersect face f;

2. tetrahedron T has three vertices in common with set s which indicates that a face of tetrahedron T forms a part or the whole of the face under consideration; and

3. tetrahedron T has four vertices in common with set s which indicates that the current triangulation is invalid since the tetrahedron vertices are coplanar.

During this checking phase all tetrahedra which have three shared vertices with set s are counted, while all tetrahedra which may intersect face f are stored. After all the tetrahedra from the ball have been examined a check is performed on whether the face is recovered. The face is recovered if the Euler equation is satisfied:

$$\mathtt{Nt} = 2(\mathtt{Nn} - 1) - \mathtt{Nh} \tag{7.12}$$

where Nt is the number of faces (triangles) which have three shared vertices with set s and which partially recover the face f, Nn is the number of points on face f (in set s), Nh is the number of points on the face edges. If the Euler equation is not satisfied it means that some edges of the stored tetrahedra intersect face f. In this case the actual intersection points are determined and they are added to the intersecting edges by means of the algorithm described in Section 7.10.2 and at the same time they are added to the set of points on face f (set s). Once all intersection points are determined the algorithm again starts to examine the ball of all points in set s.

7.11 Recovery of the mesh domain

The triangulated domain is enclosed by the cuboid. Once all boundaries of the domain have been recovered it is possible to remove elements from outside the domain. First, the procedure marks all tetrahedra which share a vertex with the cuboid as an element being outside the mesh domain. The faces of these marked tetrahedra are checked to determine whether they are boundary faces of the mesh domain. If the face is not a boundary face the neighbouring tetrahedron is examined. If the neighbouring tetrahedron is not already marked as an element being outside, then it is marked and it is used to explore further neighbouring elements. When there are no more neighbouring element which are outside the mesh domain the procedure terminates and the marked tetrahedra are removed.

7.12 Post-processing of the mesh domain

The quality of the generated mesh can be improved by some post-processing algorithms. However these algorithms do not preserve the Delaunay property of the mesh. A more detailed description of the implemented post-processing algorithms can be found in Chapter 9.

7.13 Concluding remarks

In this chapter the Delaunay principles have been expanded from two to three dimensions. The implementation of a tetrahedral mesh generation has ben discussed. Example meshes will be given in Chapter 10.

Bibliography

[1] Badouel, D., *Graphics Gems*, Chapter "Ray Tracing", pages 390–393, Academic Press, Inc., 1990.

[2] Borouchaki, H., George, P.L., Lo, S.H., "Optimal Delaunay Point Insertion", *International Journal for Numerical Methods in Engineering*, 39: 3407–3437, 1996.

[3] Bowyer, A., "Computing Dirichlet tessellations", *The Computer Journal*, 24(2): 162–166, 1981.

[4] George, P.L., Borouchaki, H., *Delaunay Triangulation and Meshing, Application to Finite Elements*, Editions Hermes, Paris, France, 1998.

[5] Haines, E., Moller, T., "Triangle Intersection Test", *Dr. Dobb's Journal*, pages 32–36, aug 2000.

[6] Joe, B., "Construction of Three-Dimensional Triangulations using Local Transformations", *Computer Aided Geometric Design*, 8: 123–142, 1991.

[7] Li, X., *Sliver-free three dimensional Delaunay mesh generation*, PhD thesis, University of Illinois at Urbana-Champaign, 2000.

[8] Shewchuk, J.R., "Adaptive Precision Floating-Point Arithmetic and Fast Robust", Technical Report CMU-CS-96-140, School of Computer Science, Carnegie Mellon University, 1996.

[9] Shewchuk, J.R., *Delaunay Refinement Mesh Generation*, PhD thesis, Carnegie Mellon University, Pittsburgh, PA, USA, 1997.

[10] Voorhies, D., Kirk, D., *Graphics Gems II*, Chapter "Ray Tracing", pages 257–263, Academic Press, Inc., 1991.

[11] Watson, D.F., "Computing the n-dimensional Delaunay tesselation with application to Voronoi polytopes", *The Computer Journal*, 24(2): 167–172, 1981.

[12] Weatherill, N.P., "The reconstruction of boundary contours and surfaces in arbitrary unstructured triangular and tetrahedral grids", *Engineering Computations*, 13(8): 66–81, 1996.

[13] Weisstein, E.W., *The CRC Concise Encyclopedia of Mathematics*, CRC Press, Boca Raton, USA, 1998.

Chapter 8

Element transformation

8.1 Introduction

In this chapter some automated procedures for transforming meshes with triangular elements into a mesh with quadrilateral elements are described. The basic concept for transformation of triangular elements into quadrilateral elements is to combine a pair of triangles that share a common side. A problem with this method is that it may result in isolated triangular elements which have no partner with which to combine. In order to overcome these isolated islands of lone triangles two different approaches were introduced by Johnston *et al* [3] and Rank *et al* [4]. One of the most popular and most widely used approaches is the Rank *et al* method.

8.2 The Rank *et al* method

The Rank *et al* method [4] comprises two strategies to transform triangles into quadrilaterals. Both are extremely easy to implement and can be combined to form a simple and efficient algorithm. This method has been widely used in engineering and graphical procedures and has been shown to be very robust.

In the first step as many triangles as possible are combined into quadrilateral elements. Figure 8.1 shows an example mesh with triangular elements and Figure 8.2 shows the same mesh after the triangles have been paired as quadrilaterals. After this step there might be some isolated islands of single triangular elements. At this point two basic strategies should be applied to convert the mesh into a fully quadrilateral mesh as proposed by Rank *et al* [4]. The first strategy is that pairs of triangular elements may be split into four smaller quadrilateral elements as shown in Figure 8.3.

The second strategy provides the solution for the triangle islands where a single triangle may be split into three quadrilaterals, connecting points on the sides of the triangle with an interior point as shown in Figure 8.4. Obviously it is possible to transform any triangular mesh into a quadrilateral mesh using this technique above.

The order of the combination and splitting steps is important to ensure the 'smoothness' of the converted meshes. Figure 8.1 shows a uniform unstructured triangular mesh. In Figure 8.2 pairs of triangles are added together to form quadrilateral elements. This leave isolated single triangular elements. Finally as shown in Figure 8.5, the quadrilateral elements are split into four quadrilaterals and the triangles into three quadrilaterals. These figures demonstrate the Rank *et al* procedure on a small mesh.

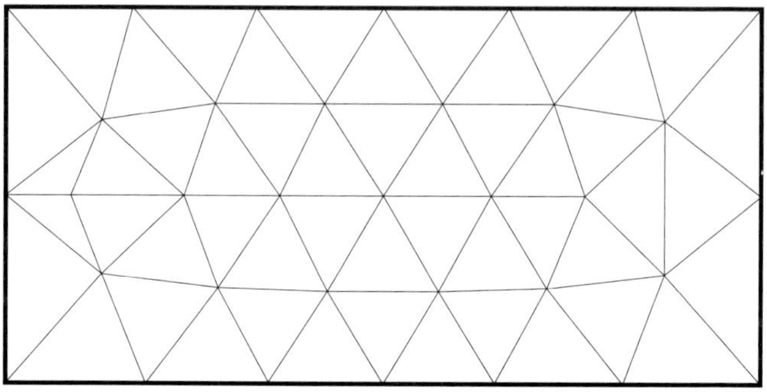

Figure 8.1: The original triangular mesh.

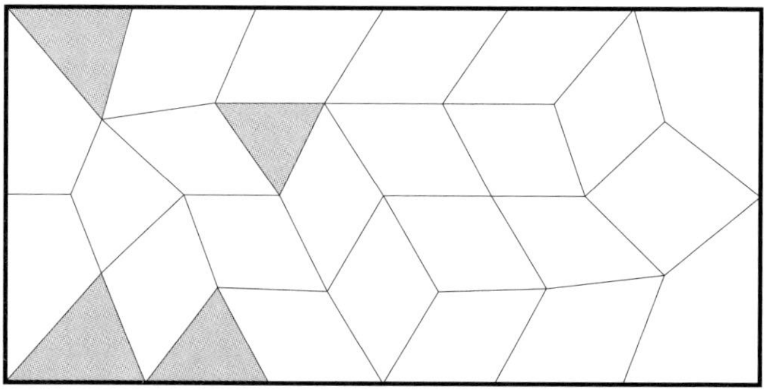

Figure 8.2: Isolated islands left in the mesh after pairing the triangles.

8.3 Mesh transformation using fission elements

The basic concept for this transformation strategy is the use of fission elements to eliminate the possibility of creating isolated triangles in the first place, rather than removing them after they have been generated.

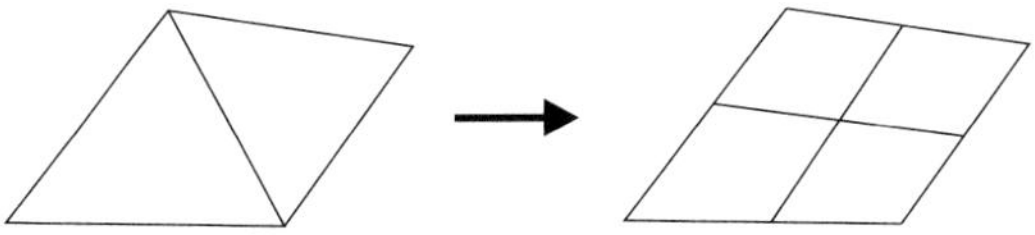

Figure 8.3: First strategy: Two triangles are combined to a quadrilateral element, then split to four quadrilateral elements.

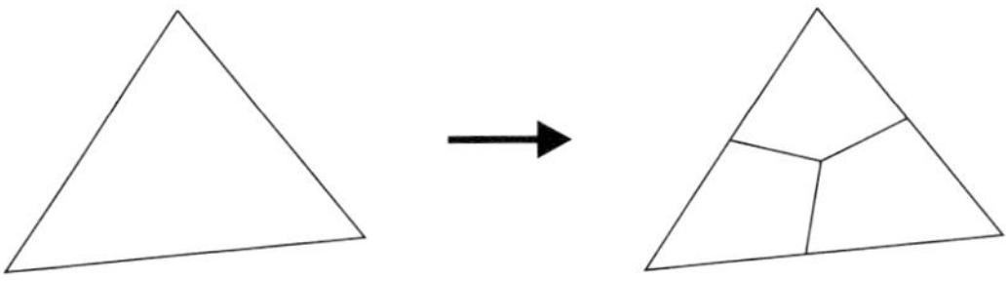

Figure 8.4: Second strategy: A triangle is split to three quadrilateral elements.

8.3.1 The fission element

The fission element was developed for quadrilateral mesh generation and was first described in reference [1]. An improved sequential and parallel quadrilateral mesh generation technique using such elements was introduced in reference [2]. The fission element is basically a trianglular element which is divided into three further trianglular elements by inserting a node at the centre of the element. Figure 8.6 shows an example for the use of the fission element. In Figure 8.6(a) the combination of $\triangle_{abc}$ and $\triangle_{bdc}$ would generate an island triangle element $\triangle_{cde}$ and an odd number of elements in the remaining subregion. The elements of the subregion are shown shaded in Figure 8.6(b). To solve the problem in a simple and efficient way a fission element is created, which means that point n is inserted in the triangle $\triangle_{bdc}$ at the centre as shown in Figure 8.6(c). In this way two quadrilateral elements, $abnc$ and $decn$, can be formed immediately (see Figure 8.6(d) and the remaining number of triangles is even in the subregion. Figure 8.6(e) shows the final step in the merging process.

8.3.1.1 The transformation procedure

Zhu *et al* [5] established a quadrilateral mesh generation method where two triangles are generated in an advancing front fashion and they are combined together into quadrilateral elements. Cheng and Topping [1] improved this approach as the fission element method for post-processing of triangular meshes.

The approach of merging pairs of elements may be implemented in a number of ways. If pairs of elements are gradually merged commencing at the first numbered triangle of the mesh the following situations may be encountered:

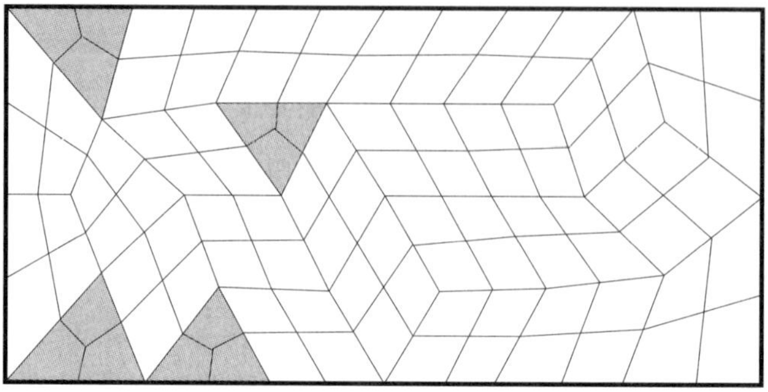

Figure 8.5: The final quadrilateral mesh.

1. If no subregion has been formed by the selection of the first triangle, then any side which is active in the selected triangle may be used to select the second triangle. A side is active if it belongs to a triangular element other than the selected triangle. When both triangular elements that share a side have been merged into a quadrilateral element then the side is flagged as inactive.

2. If two subregions are formed by the selection of the first triangle, and one of the subregions is empty, i.e. all the segments of the subregion are inactive or in other words all triangular elements are converted to quadrilateral elements in the subregion, then the remaining active side in the first triangle must be selected as the base for the second triangle selection.

3. If both subregions contain active sides then a check is carried out to determine whether the number of the remaining triangles in the subregions is even or odd. Figure 8.7 shows a situation where the already converted elements are shown by light grey shading and the first selected triangle is shaded by dark grey. The boundaries of the two subregions are indicated by dashed lines. If the number of the active triangles is even in all subregions then the merge of the two triangles is permitted. When the number of active triangles in the subregions is odd, a fission element should be inserted. This will ensure that the number of active triangles in all subregions is always even.

The sequence of the combination of triangles has been found to play a very important role in this transformation technique. The combination more usually starts with the elements with the smallest number of neighbouring elements, in other words, with the elements along the boundary. The method moves towards the interior area as with the advancing front method. The transformation terminates when there is no trianglular element left inside the domain. This technique is referred to as the advancing-element strategy.

This advancing-element strategy and the quality checks used for the quadrilateral

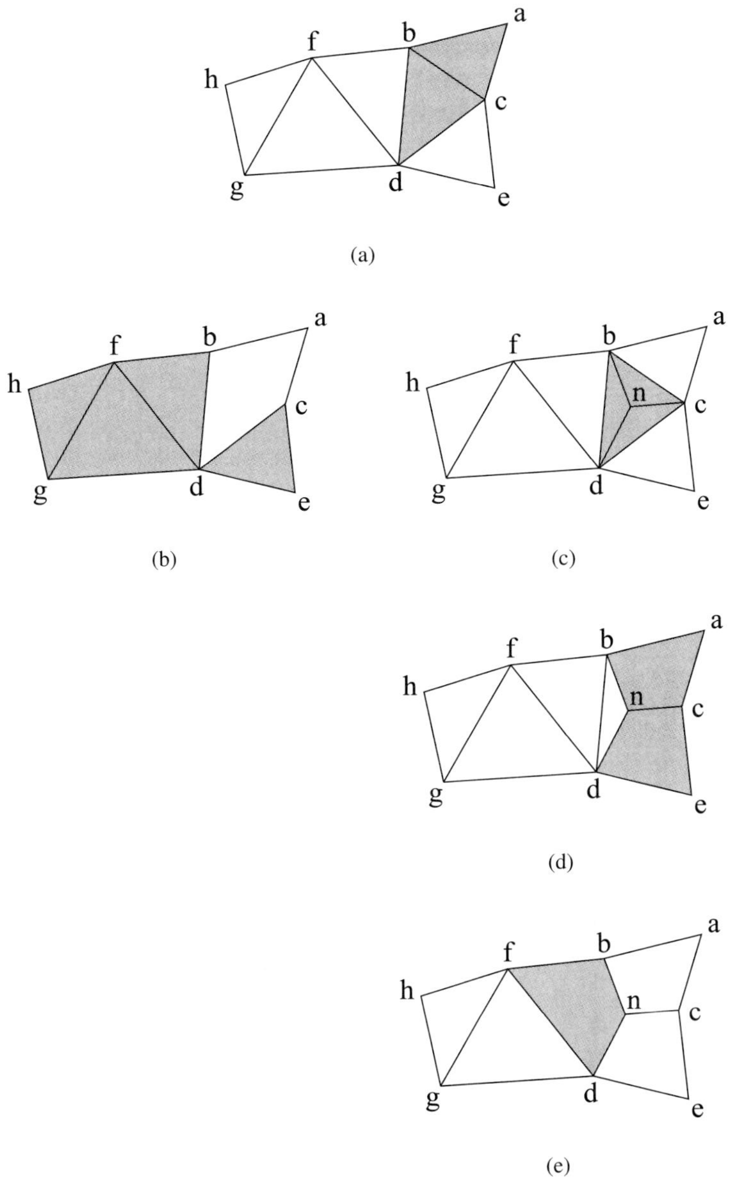

Figure 8.6: The use of fission elements to form quadrilateral elements.

elements formed makes the combination sequence of triangles both simple and efficient. It reduces the number of fission elements required for eliminating the isolated

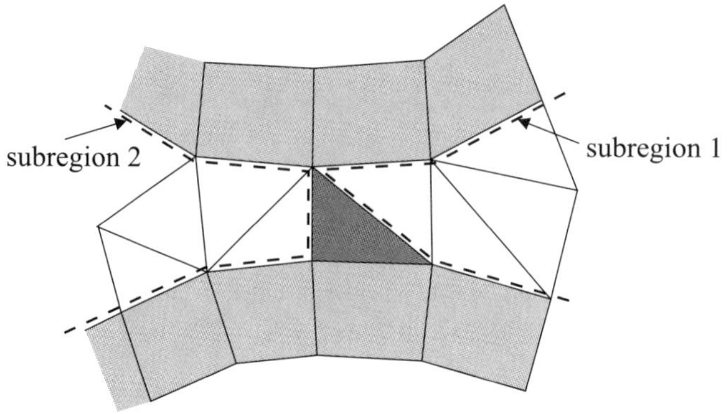

Figure 8.7: Two subregions are formed by triangle selection.

islands and produces high quality quadrilaterals.

8.3.1.2 Implementation of the transformation

The transformation procedure commences with the determination of essential data such as the element adjacency matrix of the original triangular mesh. Two data structures are defined:

1. *ElemDegree:*
   an integer array containing the number of elements adjacent to each element. Two elements are adjacent if they share an interface, in this case a side.

2. *ElemAdjac:*
   an integer matrix containing the indices of the elements adjacent to each element.

These two sets of data store the element connectivity information during combination and are modified continuously as new quadrilaterals are generated. The main steps of the transformation are:

1. **The empty domain check**:
   Check if the number of triangles in the domain is zero. Terminate the transformation if this is the case, otherwise go to the next stage.

2. **The determination of two triangles for combination**:
   Find the elements which have the smallest degree of adjacency and mark them as the candidate elements for the first pair of triangles to combine.

3. **Element quality check**:
   Assess the quality of the quadrilaterals formed by the candidate triangles and their adjacent elements. Select the pair which gives the best quadrilateral.

   The assessment may involve:

   - the determination of the largest and the smallest interior angles of the quadrilateral element and
   - a comparison between the two diagonal lengths of the quadrilateral element after combination of the two triangles.

   Element quality will be further discussed in Chapter 9.

4. **Subdomain check**:
   After selecting two triangles for combination check if any subregions are formed as a result of the triangle combination. If this is the case check if the number of active triangles in each subregion is odd.

   If the number of active triangles in a subregion is odd a fission element should be created to make the number of triangles in each subdomain even.

5. **Updating the two data structures**:
   The two triangles which are combined should be marked as inactive and *ElemDegree* and *ElemAdjac* are updated.

6. Go back to stage one.

8.4 Concluding remarks

In this chapter methods for converting triangular meshes to quadrilateral meshes are described. In the next chapter post-processing of both triangular and quadrilateral meshes will be reviewed.

Bibliography

[1] Cheng, B., Topping, B.H.V., "Improved Quadrilateral Adaptive Mesh Generation using Fission Elements", in Topping, B.H.V., Editor, *Developments in Computational Technology for Structural Engineering*, pages 391–401. Civil-Comp Press, Edinburgh, UK, 1995.

[2] Cheng, B., Topping, B.H.V., "Parallel Quadrilateral Mesh Generation", in Topping, B.H.V., Editor, *Advances in Computational Structures Technology*, pages 313–326. Civil-Comp Press, Edinburgh, UK, 1996.

[3] Johnston, B.P., Sullivan, J.M., Kwasnik, A., "Automatic conversion of triangular element meshes to quadrilateral elements", *International Journal for Numerical Methods in Engineering*, 31: 67–84, 1991.

[4] Rank, E., Schweingeruber, M., Sommer, M., "Adaptive mesh generation and transformation of triangular to quadrilateral meshes", *Commmunications in Numerical Methods in Engineering*, 9: 121–129, 1993.

[5] Zhu, J.Z., Zienkiewicz, O.C., Hinton, E., Wu, J., "A new approach to the development of automatic quadrilateral mesh generation", *International Journal for Numerical Methods in Engineering*, 32: 849–866, 1991.

Chapter 9

Mesh Quality and Post-processing

9.1 Introduction

For the post-processing algorithms mesh quality measures must be established as they govern the post-processing process. The mesh quality measures are discussed in Section 9.2. A useful overview of most of the available post-processing and smoothing techniques can be found in Canann *et al* [3]. These techniques may be categorised as:

- topological operators:
  - node insertion or local refinement techniques,
  - edge/face swaps, and
  - node or element removal
- smoothing, where the position of the nodes is changed to improve the quality of the elements in the mesh without modifying the mesh connectivity.
  - Laplacian smoothing,
  - optimization based smoothing, and
  - physics based smoothing.

9.2 Mesh quality

Element quality indicators or shape measures can be used to construct a global indicator of the mesh quality of the complete mesh. However different authors give different recommendations as to how that global indicator should be constructed. In references [10, 15] the authors advocate that one unacceptable element renders the mesh "useless" (especially in CFD computations) and propose to use the global minimum of the element shape measure as the mesh quality indicator. This is contrary to reference [6] where the authors suggest the use of a geometric average of the element shape measure or the normalized standard deviation of the edge lengths as the mesh quality indicator for tetrahedral meshes. Mesh quality is discussed in reference [1] with respect to error estimates where Berzins states that generally the mesh quality should be discussed in relation to the numerical solution technique that is to be used on the mesh. Certain numerical schemes require special properties from the mesh. However, here, a more practical approach is considered and the quality of the meshes is discussed only through certain geometric properties.

9.3 Element shape measures

It will be assumed that an element shape measure is a continuous strictly monotonic function, which is invariant under translation, rotation, reflection and uniform scaling. This definition is given for tetrahedral meshes in reference [13]. The function reaches its global maximum for "regular" shape and its global minimum for any degenerate cases. The regular shape is an equilateral triangle, an equal sided square or an equilateral tetrahedron. For convenience the shape measure is often scaled into the range of $[0, 1]$ with value 1 being the measure of the regular shape. It is possible to construct many functions obeying the above definition. Some of the most common shape measures presented in this book can be applied to isotropic meshes. For a discussion of shape measures for anisotropic triangular or tetrahedral meshes see references [9, 10].

9.4 Shape measures for triangular elements

Several shape measures exist for triangular elements, for example in references [10, 14]. In this section some of the most basic shape measures are discussed which are either related to the edge length or the internal angle of the triangular element.

- **The element aspect ratio** as defined by

$$Q = \alpha \frac{h_{max}}{\rho} \tag{9.1}$$

  where h_{max} is the longest edge of the triangle and ρ is the radius of the inscribed circle. ($\alpha = \frac{\sqrt{3}}{6}$ is a normalisation factor to ensure that an equilateral triangle has a value $Q = 1$.)

- **The length ratio** which is the ratio of the smallest side length to the average side length.

The quality of a uniform mesh in general can be assessed by criteria, proposed in reference [5], such as:

- **The standard deviation of angles** as defined by

$$\sigma_a = \sqrt{\sum_{i=1}^{n_a} \frac{(\alpha_i - 60^o)^2}{n_a}} \tag{9.2}$$

where α_i stands for all the angles in the triangulation, n_a being equal to three times the number of triangles.

- **The standard deviation of edge length** as defined by

$$\sigma_l = \sqrt{\sum_{i=1}^{n_t} \frac{\left(L_i^1 - \bar{L}_i\right)^2 + \left(L_i^2 - \bar{L}_i\right)^2 + \left(L_i^3 - \bar{L}_i\right)^2}{3n_t}} \tag{9.3}$$

where L_i^1, L_i^2 and L_i^3 are the lengths of each triangle, their average being denoted by $\bar{L}_i$ and n_t being the number of triangles.

9.5 Shape measures for quadrilateral elements

In the case of quadrilateral elements quality measures have been described in Section 5.4.2. Equation 5.1 also satisfies the definition of an element shape measure function as given in Section 9.3. However some other shape measures are also available for quadrilateral elements (for example see reference [2]).

skew - $\max(\cos(\alpha))$ where α is the angle between the edges of the element as shown in Figure 9.1b.

taper -

$$\tau = 4\,\frac{\min(A_1, A_2, A_3, A_4)}{A} \tag{9.4}$$

where A is the area of the quadrilateral element, and A_i are the areas of the triangles formed by the centre of the quadrilateral element and by two adjacent vertices as shown in Figure 9.1c. The value of τ is one for the ideal shape and it tends to zero as the shape deforms.

minimum angle - The minimum angle is the smallest internal angle.

maximum angle - The maximum angle is the largest internal angle.

9.6 Tetrahedron element shape measures

In the case of two dimensional meshes visualisation is helpful in pointing out discrepancies in the mesh. In three dimensions this is a rather difficult task. In three dimensions a more rigorous approach is taken to determine the quality of the mesh.

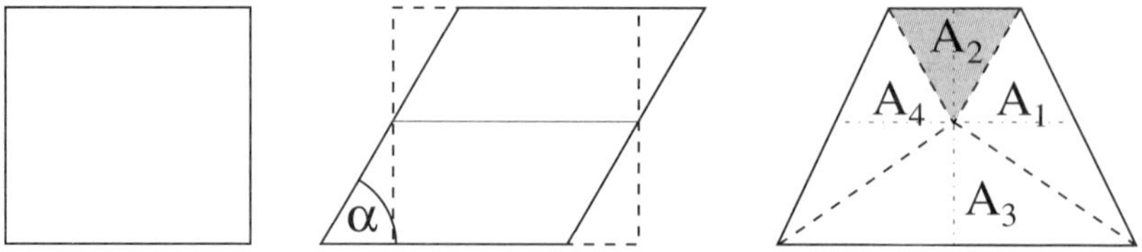

(a) Ideal shape (b) Skew (c) Taper

Figure 9.1: Quadrilateral element shapes.

9.6.1 The shape measure based on solid angle

A tetrahedron T defined by the vertices $\{t_0, t_1, t_2, t_3\}$ is to be considered. If the points of the face opposite vertex t_i are projected on the unit sphere centred at the vertex t_i, then the surface area of the projected points is a measure of the solid angle at the vertex t_i of the tetrahedron. The solid angle at a vertex of a tetrahedron will be in the range $[0, 2\pi]$ (both extreme values denote a degenerate tetrahedron). It can be shown (see reference [8]) that the sum of solid angles Θ_i of a tetrahedron is $0 \leq \sum_{i=0}^{3} \Theta_i \leq 2\pi$. A tetrahedron with a large, nearly flat solid angle must also have a very small solid angle. Thus the minimum solid angle of a tetrahedron can be used as a shape measure (this obeys all other assumptions for the shape measure).

The direct calculation of a solid angle can be expensive as it requires the calculation of inverse trigonometric functions. In reference [13] Lius and Joe proposed to use a measure based on the *sine* of the half solid angle and give the following formula to calculate the sine of the half solid angle

$$\sin(\Theta_i/2) = 12v \left(\prod_{\substack{j,k \neq i \\ 0 \leq j < k \leq 3}} \left((l_{ij} + l_{ik})^2 - l_{jk}^2 \right) \right)^{-\frac{1}{2}} \tag{9.5}$$

where v is the tetrahedron volume, and l_{ij} are tetrahedron edge lengths. The measure built on the *sine* of the half solid angle is:

$$\sigma = \beta \min_i \sigma_i \qquad \text{where} \quad \sigma_i = \sin(\Theta_i/2), \quad i \in \{1, 2, 3, 4\} \tag{9.6}$$

where $\beta^{-1} = \sqrt{6}/9$ is the value of σ_i for the four solid angles of the regular tetrahedron.

9.6.2 The radius ratio

The ratio between the radius of the inscribed sphere and the radius of circumscribed sphere of a tetrahedron can also be used as a shape measure. The radius ratio is defined as

$$\rho = 3\frac{r_{\text{in}}}{r_{\text{out}}} \tag{9.7}$$

where r_{in} is the radius of the inscribed sphere and r_{out} is the radius of the circumscribed sphere. In reference [13] the following formula is given to calculate ρ:

$$r_{\text{in}} = \frac{3v}{\sum_{i=0}^{3} s_i} \tag{9.8}$$

$$r_{\text{out}} = \frac{\sqrt{(a+b+c)(a+b-c)(a+c-b)(b+c-a)}}{24v} \tag{9.9}$$

$$\rho = 3\frac{r_{\text{in}}}{r_{\text{out}}} = \frac{216v^2}{\sqrt{(a+b+c)(a+b-c)(a+c-b)(b+c-a)}\sum_{i=0}^{3} s_i} \tag{9.10}$$

where v is the tetrahedron volume, a,b and c are the products of the lengths of opposite edges, and s_i are the face areas.

9.6.3 The mean ratio

In reference [12] Liu and Joe introduced a new shape measure called the mean ratio.

Let $T(t_0, t_1, t_2, t_3)$ be any non-degenerate tetrahedron and $R(r_0, r_1, r_2, r_3)$ be a regular tetrahedron with the same volume as T. Let $\mathbf{M}$ be the matrix involved in an affine transformation from R to T, i.e. $\mathbf{t} = \mathbf{M}\mathbf{r}+\mathbf{b}$ where $\mathbf{b}$ is a translation vector (matrix $\mathbf{M}$ also depends on the ordering of vertices in T and R). The mean ratio η of tetrahedron T is defined as:

$$\eta(T) = \frac{3\sqrt[3]{\lambda_1\lambda_2\lambda_3}}{\lambda_1+\lambda_2+\lambda_3} \tag{9.11}$$

where λ_1, λ_2, λ_3 are the eigenvalues of matrix $\mathbf{A} = \mathbf{M}^T\mathbf{M}$. The eigenvalues λ_1, λ_2, λ_3 are positive as eigenvalues of a symmetric matrix and $0 < \eta(T) \leq 1$. The value of $\eta(T)$ tends to 0 as T becomes a degenerate tetrahedron. If T is a regular tetrahedron then $\eta(T) = 1$. Reference [13] also gives a simple formula to calculate $\eta(T)$:

$$\eta(T) = \frac{12(3v)^{\frac{2}{3}}}{\sum\limits_{0\leq i<j\leq 3} l_{ij}^2} \tag{9.12}$$

where v is the tetrahedron volume and l_{ij} are the tetrahedron edge lengths.

9.6.4 Other shape measures

Besides the shape measures described above, reference [6] defines a number of shape measures called aspect ratios. These measures are built as a ratio of some quantity that vanishes with volume approaching zero over another quantity which does not vanish when the volume reaches zero. In particular the numerator of an aspect ratio may be chosen to be proportional to the volume of the tetrahedron, the radius, area (or volume) of an inscribed sphere, or the minimum solid angle. The denominator can be chosen as the length of the longest edge, the average of edge lengths, the sum of edge lengths, the area or volume of the circumsphere, etc. The numerator of the aspect ratio cannot be chosen as the function of circumsphere, or the smallest edge or minimum dihedral angle because degenerate tetrahedron may have non-null circumsphere, smallest edge or dihedral angle.

As an example of the aspect ratio shape measure reference [6] points out that the γ aspect ratio may be defined as

$$\gamma = \frac{12}{\sqrt{6}} \frac{r_{\text{in}}}{\max\limits_{0 \le i < j \le 3} l_{ij}} \tag{9.13}$$

9.6.5 Pseudo shape measures

There are some quantities which are not true shape measures in the sense of the definition given in section 9.3 but which are often used to evaluate the shape of a tetrahedron. Reference [6] mentions two of them:

- the minimum dihedral angle
- the edge ratio

9.6.5.1 The minimum dihedral angle

The dihedral angle is an angle between two planes. If two planes intersect along edge E then the dihedral angle between the two planes is the angle between the intersection of the planes with a third plane constructed perpendicular to the edge E. In the case of a tetrahedron one can distinguish six dihedral angles, one dihedral angle for each pair of planes intersecting at the tetrahedron edges.

With the convention that the signed volume of the tetrahedron is positive when its face normals point outside the tetrahedron, the dihedral angles are bound between 0 and π. They can be calculated as π minus the angle between face normals. The minimum dihedral angle $\phi_{\min}$ can be used as a shape measure:

$$\phi_{\min} = \alpha \min_{0 \le i < j \le 3} \phi_{ij} = \alpha \min_{0 \le i < j \le 3} (\pi - \arccos(\mathbf{n}_{ij1} \cdot \mathbf{n}_{ij2})) \tag{9.14}$$

where $\mathbf{n}_{ij1}$ and $\mathbf{n}_{ij2}$ are the normals of the two faces adjacent to the edge ij and $\alpha^{-1} = \pi - \arccos(2\pi/3)$ is the value of all dihedral angles of the regular tetrahedron.

However, the measure described above violates, the assumption that the shape measure reaches 0 for degenerate tetrahedra. In the case of a tetrahedron called 'thorn' or 'needle' the volume of such a tetrahedron can be made arbitrarily small but the minimal dihedral angle can still be non-zero. The needle tetrahedron $T'(t'_0, t'_1, t'_2, t'_3)$

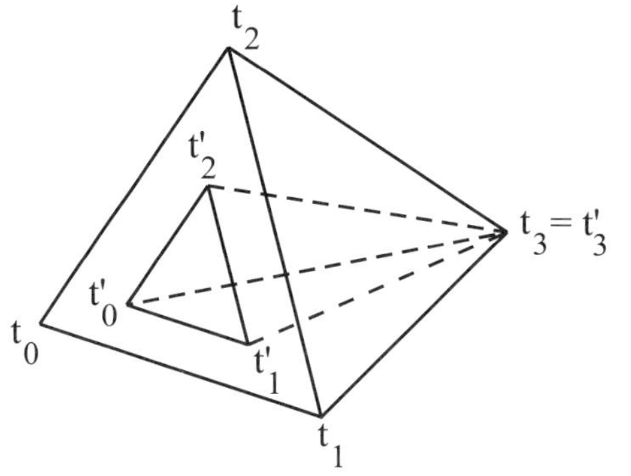

Figure 9.2: A needle tetrahedron.

shown in Figure 9.2 can be made by taking a regular tetrahedron $T(t_0, t_1, t_2, t_3)$ and uniformly moving the vertices t_0, t_1, t_2 towards the centroid of the face $t_0t_1t_2$. In this way the volume of the resulting tetrahedron can be made arbitrarily close to zero while the dihedral angles at edges $t'_0t'_1$, $t'_1t'_2$, $t'_2t'_0$ stay close to $90°$ and the dihedral angles at edges $t'_0t'_3$, $t'_0t'_3$,$t'_2t'_3$ stay close to $60°$.

9.6.5.2 The edge ratio

The ratio of the shortest edge length over the longest edge length of the teterahedron is called the edge ratio:

$$r = \frac{\min\limits_{0 \le i < j \le 3} l_{ij}}{\max\limits_{0 \le i < j \le 3} l_{ij}} \tag{9.15}$$

As in the case of the dihedral angle there is a type of degenerate tetrahedra for which the edge ratio is non-zero. Such tetrahedra are called **slivers**. The sliver tetrahedron $T'(t'_0, t'_1, t'_2, t'_3)$ shown in Figure 9.3 can be obtained by equally spacing four vertices on an equator of a sphere and then slightly displacing the vertices out of the equator. In this way the volume of T' can be made arbitrarily small but the edge lengths will not be zero.

9.6.6 Equivalence of shape measures

With so many available shape measures there arises a question about the relationship between them. In reference [13] Liu and Joe introduce the notion of the shape measure equivalence and they also prove the equivalence of the shape measures σ, ρ, η. The authors of reference [6] conjecture that all shape measures that satisfy the definition of the shape measure given in section 9.6 are equivalent.

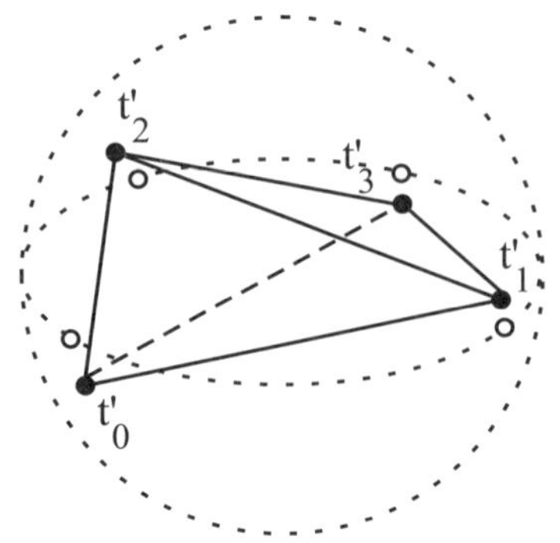

Figure 9.3: A sliver tetrahedron.

The equivalent shape measures can detect all degenerate tetrahedra types, therefore any of the shape measures can be used in mesh optimization. However, as shown in reference [13] different shape measures produce different distributions and in the case of the shape measures of σ, ρ and η, it is possible to find tetrahedra for which the ordering of those shape measures from the highest to the lowest values will take all six possible permutations.

9.7 Post-Processing: Topological operations

The topological operations change the topology of the mesh, for example the connectivity of the elements. These operations must ensure, that their application does not invalidate the mesh and they must also improve the quality of the mesh.

9.7.1 Triangular elements

9.7.1.1 Removing element triplets

Removing element triplets is basically a point removal enhancement technique. Any node with three elements connected to it forms a candidate triplet for removal. In Figure 9.4(a) three triangles abd_Δ, bcd_Δ and cad_Δ are crowded around point d. Assuming that all triangles satisfy the mesh size criteria then point d can be removed. In this case only one triangle abc_Δ would stay in the mesh as shown in Figure 9.4(b).

However it is important that none of the sides $\overline{ad}$, $\overline{bd}$ and $\overline{cd}$ are on a boundary because in that case no node removal can be carried out. An example is shown in Figure 9.5 where three triangle elements are attached to point d but sides $\overline{ad}$ and $\overline{bd}$ are on a boundary marked by the dashed line. In this case no removal can take place.

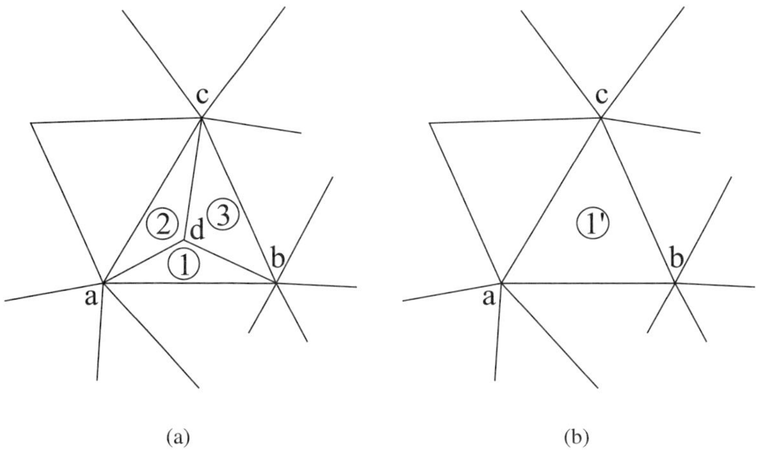

Figure 9.4: (a) Element triplet definition and (b) its removal.

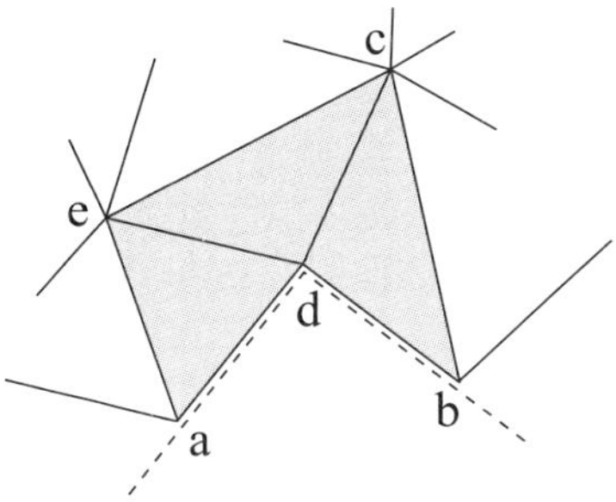

Figure 9.5: Invalid element triplet.

9.7.2 Quadrilateral elements

9.7.2.1 Element merging at interior nodes with two attached quadrilateral elements

Figure 9.6 shows a pair of quadrilateral elements that have two connected sides. This situation may occur inside a mesh and can be cleaned by merging the two quadrilateral elements into one element. The interior node is deleted. The resulting quadrilateral element will still satisfy the required side length constraints.

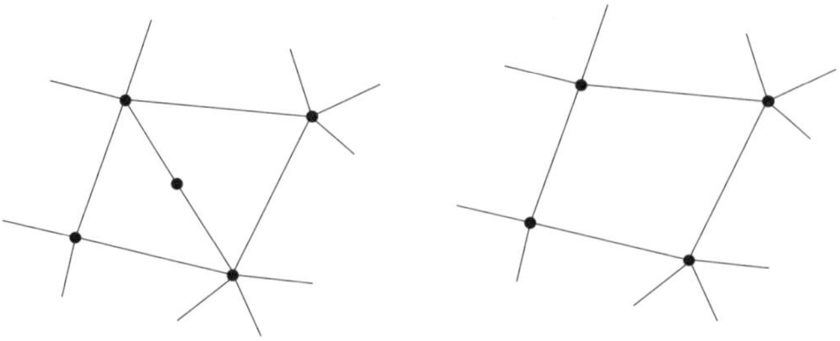

Figure 9.6: Element merging at mesh clean-up.

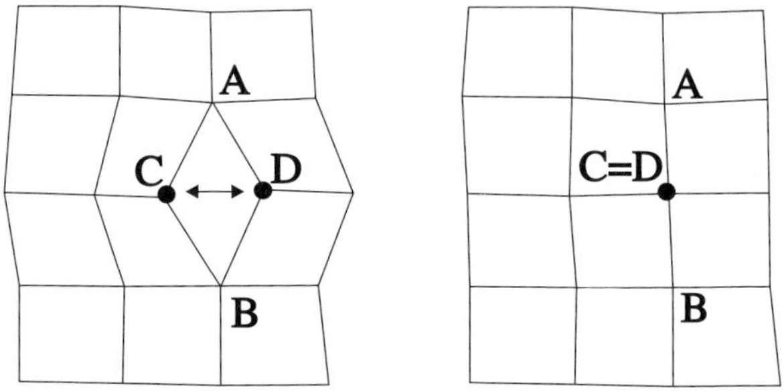

(a) Before (b) After

Figure 9.7: The element collapsing technique.

9.7.2.2 Element collapse

Figure 9.7(a) shows a configuration where quadrilateral $ACBD$ is "wedged" into the mesh. By collapsing the element, in other words merging point C and D the result is a mesh where all surrounding quadrilateral elements have the configuration with only regular nodes as shown in Figure 9.7(b). Regular nodes are discussed in Section 5.4.3 and they are shown in Figure 5.4.

9.7.2.3 Smoothing of interior nodes with three attached quadrilateral elements

At this step only interior nodes with three attached quadrilateral nodes are considered. The internal angle at the node is calculated for each attached element.

- When the maximum of the three internal angles (α) is greater than 100 degrees, then the three elements are merged into two. Figure 9.8(a) shows that this node g may be too close to the boundary of the enclosing polygon $abcdef$. The result

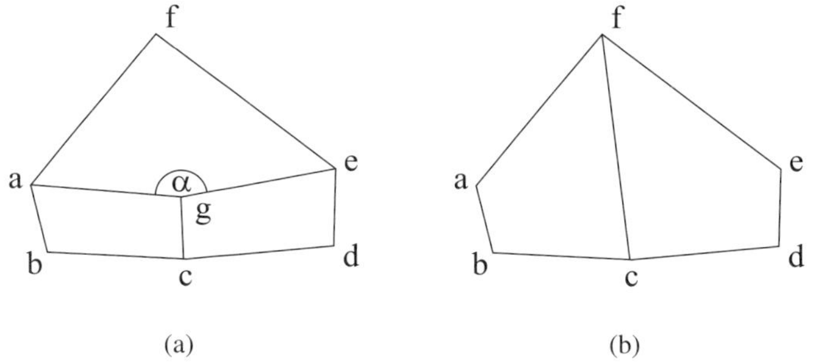

Figure 9.8: (a) Mesh enhancement when internal angle α is large and node g is too close to the boundary, (b) and the final stage.

of the combination of the three elements is shown in Figure 9.8(b). The two new quadrilateral elements can be further enhanced at the next round of smoothing.

- When the maximum internal angle (α) is less than 100 degrees, then the internal point can be repositioned in such a way that all internal angles at the point will be 120 degrees.

9.7.2.4 Enhancing boundary elements with an internal angle greater than 165 degrees

When the internal angle of a boundary quadrilateral element is greater than 165 degrees, as shown in Figure 9.9(a) at node b, the element quality can be greatly improved by using the following process [4]:

- Determine the smaller angle of $dab_{\measuredangle}$ or $dcb_{\measuredangle}$.
- Select the adjacent element which shares the node with the smaller internal angle. In this case it will be element $adfe$. Consider the area of the two quadrilaterals $abcd$ and $adfe$ as a whole. (See Figure 9.9(b).)
- Generate two new nodes h and k. There are several ways to do this. In this case node h is the middle point between node a and d, while node k is the centre of $\triangle_{ahf}$. The nodes are shown in Figure 9.9(c).
- Delete elements $abcd$ and $adfe$.
- Create four new elements: $abhk$, $bcdh$, $dfkh$ and $feak$ as shown in Figure 9.9(c).

Using this procedure the internal angle $abc_{\measuredangle}$ is split in two.

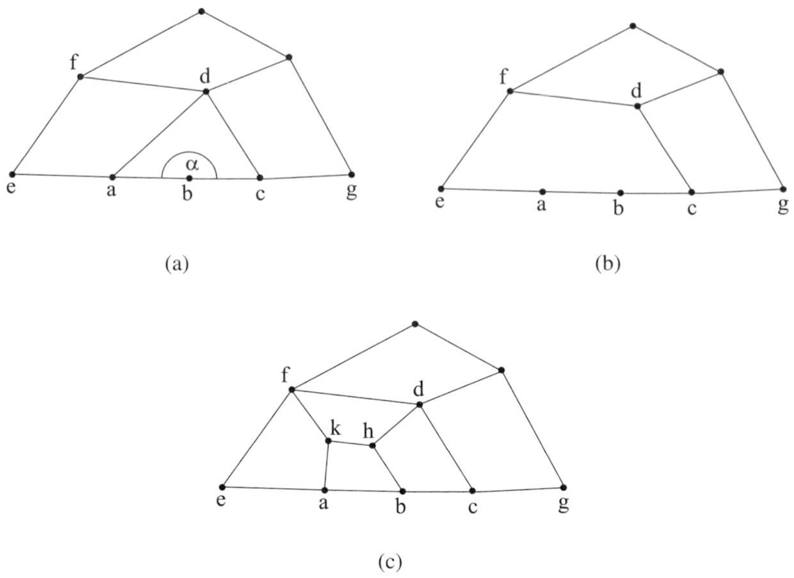

Figure 9.9: Mesh enhancement on the boundary quadrilateral elements where the internal angle (α) is greater than 165 degrees.

9.7.3 Tetrahedral elements

9.7.3.1 Edge and face swapping

The edge and face swapping algorithm is implemented using the procedure of Freitag and Ollivier-Gooch [7]. Figure 7.17 in Chapter 7 shows the face swapping algorithm. It is possible to change two tetrahedra into three tetrahedra or *vice versa* depending on the resulting quality of the mesh. Edge swapping changes an edge shell with N tetrahedra into a new configuration. The possible combinations are shown in Figure 9.10. The figure only shows unique configurations, all other configurations can be generated by a rotation. The number of possible rotations is also shown in Figure 9.10. It can be noticed that as N increases the number of tetrahedra generated by the edge swap also increases. For example when $N = 7$, seven tetrahedra are replaced by ten tetrahedra. This is the main reason why configurations with higher N numbers are not considered, they would produce too many tetrahedra. An example for the edge swap in the case of $N = 5$ is shown in Figure 9.11.

The edge or face swapping can be carried out only when the sum of the volume of all tetrahedra in the original edge shell and the sum of the volume of all tetrahedra in the final configuration are equal. Figure 9.12 shows a configuration where edge swapping is not possible.

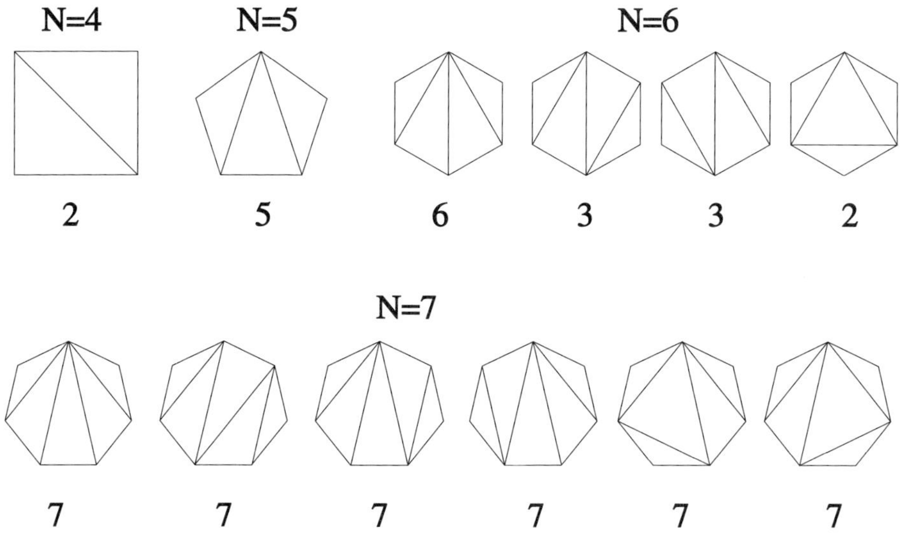

Figure 9.10: Equatorial triangles after edge swapping for $4 \geq N \geq 7$ and the number of unique rotations for each configuration.

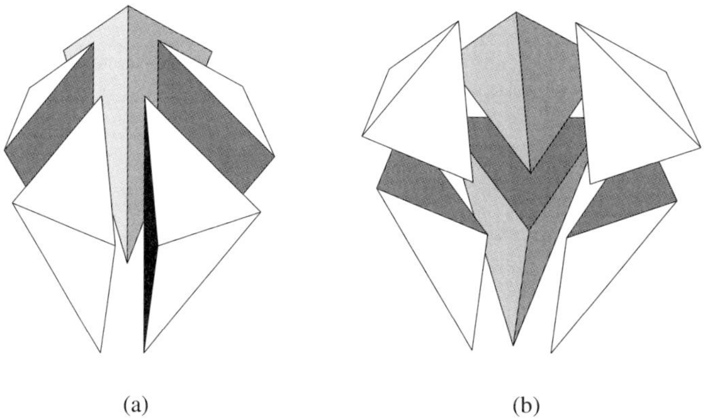

Figure 9.11: An example of edge swapping for the case when $N = 5$.

9.7.4 Sliver element removal

One of the difficult optimization steps is the removal of sliver elements, since the 3D Delaunay mesh generation does not ensure the elimination of them. A sliver is a tetrahedron whose vertices are almost coplanar with a small radius-edge ratio but arbitrarily small dihedral angle. Li [11] addressed this problem and provided theoret-

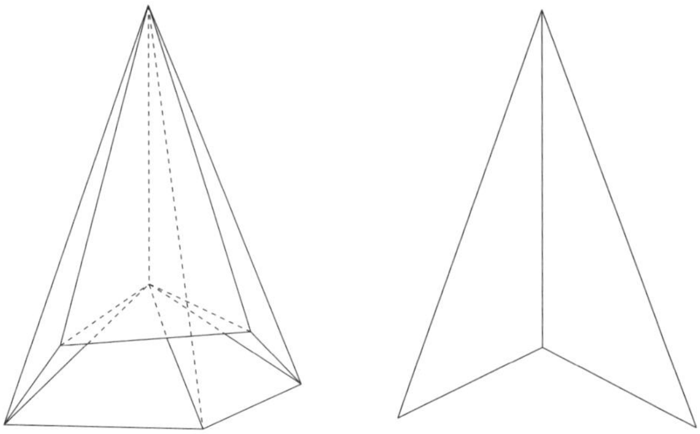

(a) 3D view

(b) 2D projected view

Figure 9.12: An example configuration where edge swapping is not possible.

ical proofs to generate sliver free meshes. In the current implementation slivers are removed by inserting a point somewhere in the circum-sphere of the sliver element. The insertion algorithm is the same as Algorithm 7.5 in Chapter 7 with the addition that a point is not selected if it would generate a new sliver with smaller circum radius than the circum radius of the sliver to be removed or the insertion quality is too low. However, it has to be noted that the current implementation of the algorithm as described in the book does not guarantee the elimination of all slivers from the mesh.

9.8 Post-Processing: Geometric operations

The geometric operations do not change the connectivity of the mesh. These operations attempt to improve the quality of the mesh by displacing the meshpoints. These operations must also ensure that their application does not invalidate the mesh.

9.8.1 Laplacian smoothing

Laplacian smoothing is one of the most common smoothing algorithms. The algorithm displaces a particular meshpoint into a new position which is determined by its neighbouring meshpoints. Neighbouring meshpoints consist of all nodes which are connected to the particular meshpoint by an edge. In its simplest form a meshpoint is placed at the centroid of the neighbouring meshpoints. This simple technique is called nodal averaging. The displacement of the meshpoints is carried out iteratively by going through all meshpoints several times. The method is very well suited to two dimensional meshes. However the algorithm can result in distorted or invalid elements in certain cases. To avoid these problems several improvements have been suggested in the literature. For example weighting the contribution of meshpoints by edge length (see Section 5.7.2 in Chapter 5) or introducing constraints with the Laplacian smoothing (see reference [3]). For the smoothing of 3D tetrahedral meshes the latter approach must be taken, where the number of improved elements and the number of elements of poorer quality are weighed against each other.

9.9 Concluding Remarks

In this chapter mesh quality assessment and mesh post-processing techniques have been discussed. In the next chapter examples of the mesh generation techniques are presented.

Bibliography

[1] Berzins, M., "Mesh Quality: A Function of Geometry, Error Estimates or Both?", in *Proceedings of 7th International Meshing Roundtable*, pages 229–237. Sandia National Labs, PO Box 5800 MS 0441, Albuquerque, NM 87185-044, October 1998, Sandia Report 98-2250.

[2] Blacker, T.D., *CUBIT Mesh Generation Environment, Volume 1: User's Manual*, Sandia National Laboratories, Albuquerque, New Mexico, 1994, SAND94-1100.

[3] Canann, S.A., Tristano, J.R., Staten, M.L., "An approach to combined Laplacian and optimization-based smoothing for triangular, quadrilateral and quad-dominant meshes", in *Proceedings of 7th International Mehing Roundtable*, 1998.

[4] Cheng, B., *Parallel adaptive finite element mesh generation*, M.Phil. Thesis, Heriot-Watt University, Edinburgh, UK, 2001.

[5] Chew, L., "Almost regular Delaunay-triangulations", *International Journal for Numerical methods in Engineering*, 40(24): 4595–4610, 1994.

[6] Dompierre, J., Labbé, P., Guibault, F., Camarero, R., "Proposal of Benchmarks for 3D Unstructured Tetrahedral Mesh Optimization", Rapport CERCA R98-91, École Polytechnique Montréal, Centre de Recherche en Calcul Appliqué 5160, boul. Décarie, suite 400, Montréal, (Québec), H3X 2H9, Canada, September 1998, Presented at 7th International Meshing Roundtable, Dearborn, Michigan, 26–28 October 1998.

[7] Freitag, L.A., Ollivier-Gooch, C.F., "Tetrahedral Mesh Improvement Using Swapping and Smoothing", *International Journal for Numerical Methods in Engineering*, 40: 3979–4002, 1997.

[8] Gaddum, J.W., "The sums of the dihedral and trihedral angles in a tetrahedron", *Amer. Math. Monthly*, 59: 370–371, 1952.

[9] George, P.L., *Automatic Mesh Generation, Application to Finite Element Methods*, John Wiley & Sons, Chichester, England, 1991.

[10] George, P.L., Borouchaki, H., *Delaunay Triangulation and Meshing, Application to Finite Elements*, Editions Hermes, Paris, France, 1998.

[11] Li, X., *Sliver-free three dimensional Delaunay mesh generation*, PhD thesis, University of Illinois at Urbana-Champaign, 2000.

[12] Liu, A., Joe, B., "On the shape of tetrahedra from bisection", *Mathematics of Computation*, 63(207): 141–154, jul 1994.

[13] Liu, A., Joe, B., "Relationshp Between Tetrahedron Shape Measures", *BIT*, 34: 268–287, 1994.

[14] Weatherill, N.P., "Unstructured grids: Procedures and Applications", in Thompson, J.F., Soni, B.K., Weatherill, N.P., Editors, *Handbook of Grid Generation*. CRC Press, Boca Raton, USA, 1999.

[15] Zavattieri, P.D., Dari, E.A., Buscaglia, G.C., "Optimization strategies in unstructured mesh generation", *International Journal for Numerical Methods in Engineering*, 39: 2055–2071, 1996.

Chapter 10

Mesh Generation Examples

10.1 Introduction

Some examples of mesh generation were presented in Chapter 1. Further small examples illustrating the mesh generation concepts and algorithms are provided throughout the chapters of this book. This chapter demonstrates some larger mesh generation tasks. Meshes created with different methods are presented and their properties compared using the mesh quality measures.

10.2 Two-dimensional meshes

10.2.1 Methods to generate triangular elements

In order to compare the methods which can be used to generate two-dimensional meshes with triangular elements the model of a wrench is meshed using four methods: advancing front, Delaunay triangulation (`CENTROID` variation, $\alpha = 1.5$, $\beta = 2.0$), structured mesh generation and paving with the quadrilateral elements split along their shortest diagonal. Figure 10.1 shows the meshes, while Table 10.1 gives an overview of the essential statistics for the different meshes. Figures 10.2 to 10.5 show the dif-

Method	No. elements	No. nodes	Relative CPU Time
Advancing front	3648	1977	75.5
Delaunay triangulation	4392	2349	1.875
Structured mesh generation	3300	1823	1
Paving with splitting	3280	1793	45.5

Table 10.1: Numbers of triangular elements and nodes in meshes generated by different methods and their comparative CPU generation time.

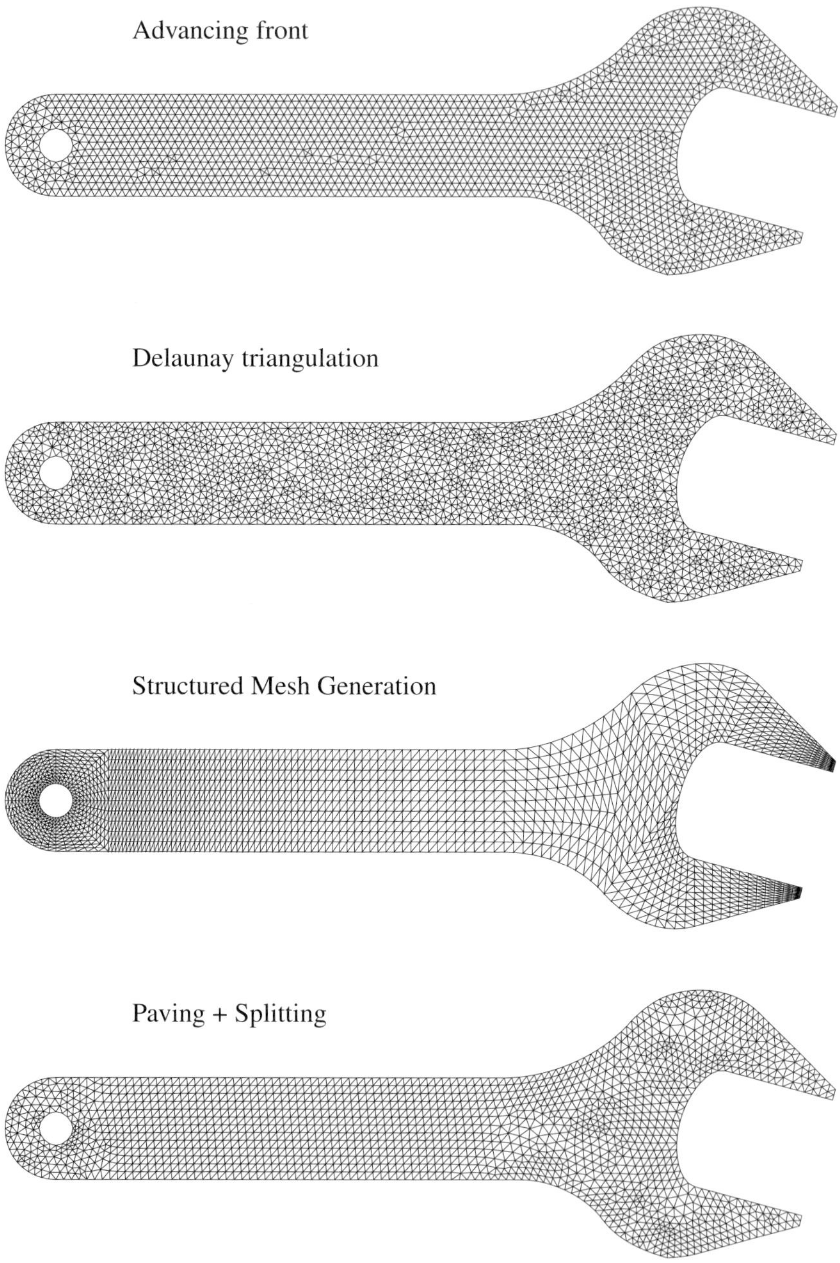

Figure 10.1: The wrench meshes.

ferent shape measures for the meshes. The results show the main differences between the advancing front and Delaunay triangulation methods. The advancing front method can generate equilateral triangles in large areas and the shape measures tend to be better for this method. On the other hand the generation of meshes using the advancing front is much slower. However the other two methods, structured mesh generation and paving with splitting, produce element qualities which are inherent to them. For example in Figure 10.5(c) the distribution of the angles is shown for the meshes generated using the structured mesh generation method. The method generates a large number of right angled triangles, which can be observed from the figure. On the other hand in Figure 10.3(c) the distribution of the length ratio shape measure for the structured mesh indicates that the triangles are stretched therefore there is no peak in the distribution at 45 degrees in Figure 10.5(c). Similarly in Figure 10.5(d) the distribution of the angles is shown for the paving method with diagonal splitting. Since paving tends to generate good quality, nearly square elements, the splitting of these elements results in right angled triangles with internal angles approximately equal to 45 or 90 degrees.

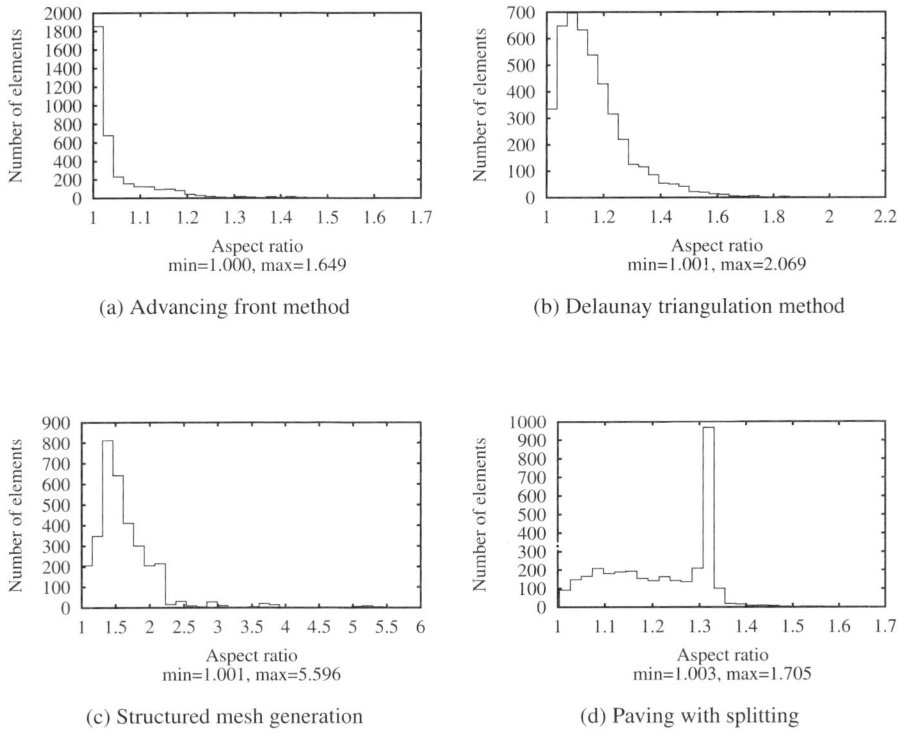

(a) Advancing front method

(b) Delaunay triangulation method

(c) Structured mesh generation

(d) Paving with splitting

Figure 10.2: Distribution of the aspect ratio shape measure as defined in Equation 9.1.

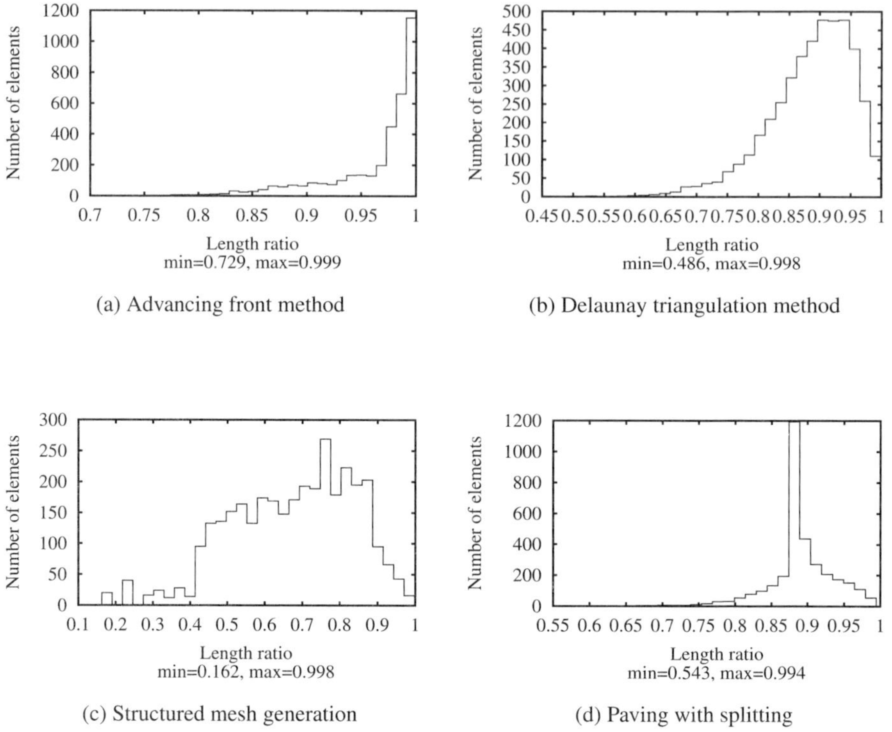

Figure 10.3: Distribution of the length ratio shape measure defined in section 9.4.

10.2.2 Methods to generate quadrilateral elements

The comparison of the methods which can generate quadrilateral elements is presented here. Four methods are considered: paving as discussed in Chapter 5 and the three mesh conversion methods discussed in Chapter 8. The conversion methods are the fission method, the Rank method and the triangle halving method. The very simple triangle halving method, not discussed in Chapter 8, is based on the technique of inserting a point into the centre of each triangle and halving the sides of the triangles in order that quadrilateral elements can be formed. This method is only provided for comparative purposes.

The domain chosen for the comparative study is a triangle shape with curved borders and a crescent moon hole within the domain. The measure for comparison is the directly generated quadrilateral mesh created by the paving method and shown in Figure 10.6a. The triangular base mesh for the conversion methods is generated using the advancing front method. However while the mesh parameter for the paving and the

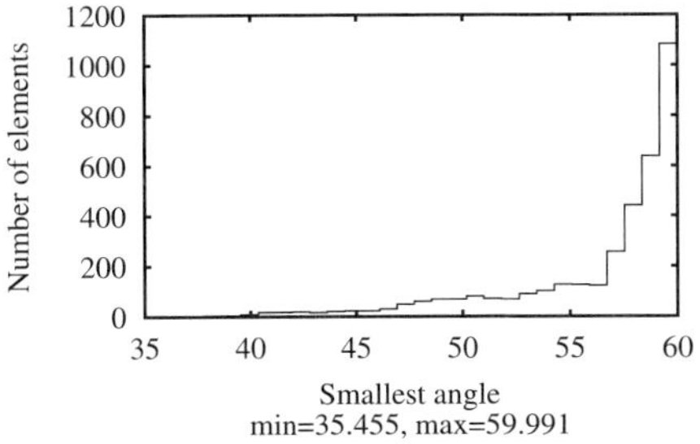

(a) Advancing front method

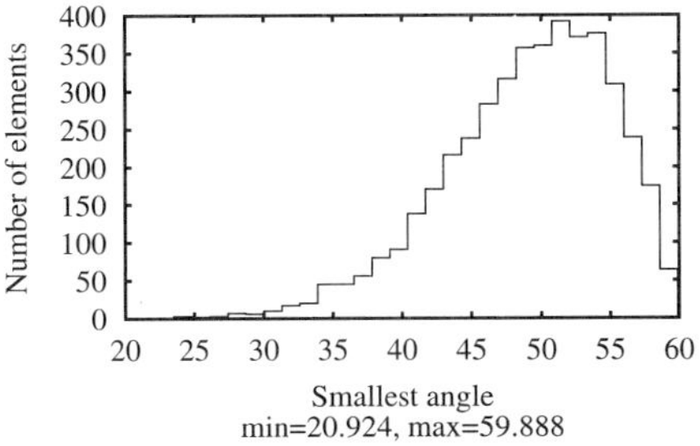

(b) Delaunay triangulation method

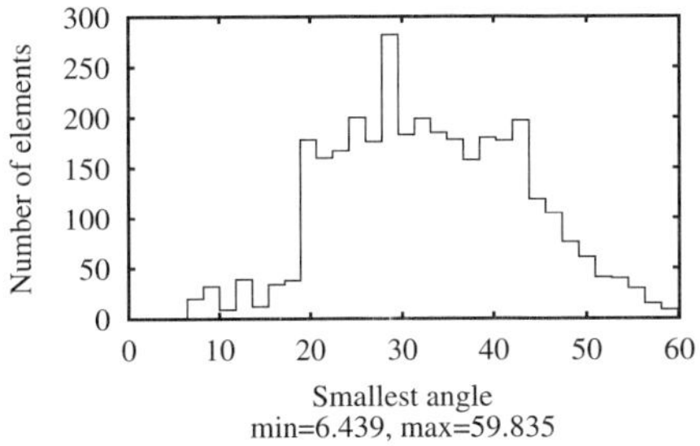

(c) Structured mesh generation

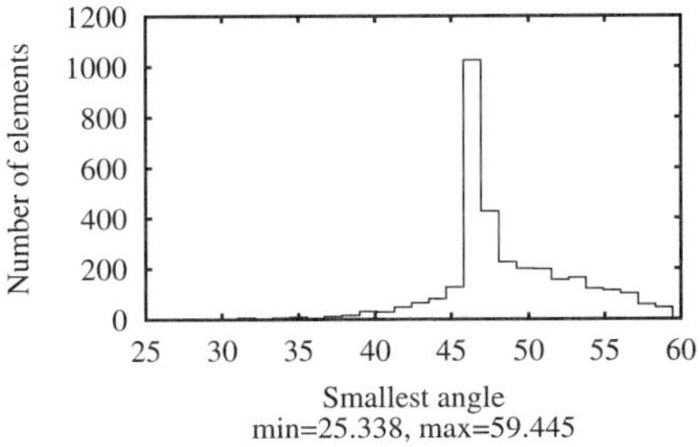

(d) Paving with splitting

Figure 10.4: Distribution of the smallest angle shape measure.

Method	No. elements	No. nodes
Paving	926	1028
Fission	1081	1183
Rank	1130	1232
Triangle halving	1686	1788

Table 10.2: Number of quadrilateral elements and nodes in the meshes generated by different methods.

fission method is δ the mesh parameter for the Rank and the triangle halving method must be $2\,\delta$; as a result of the nature of these methods that require halving the sides of the elements. In this way the same element sizes can be achieved and the results are comparable. Table 10.2 shows the number of elements and the number of nodes in the meshes. From the table it can be seen that the methods generate approximately

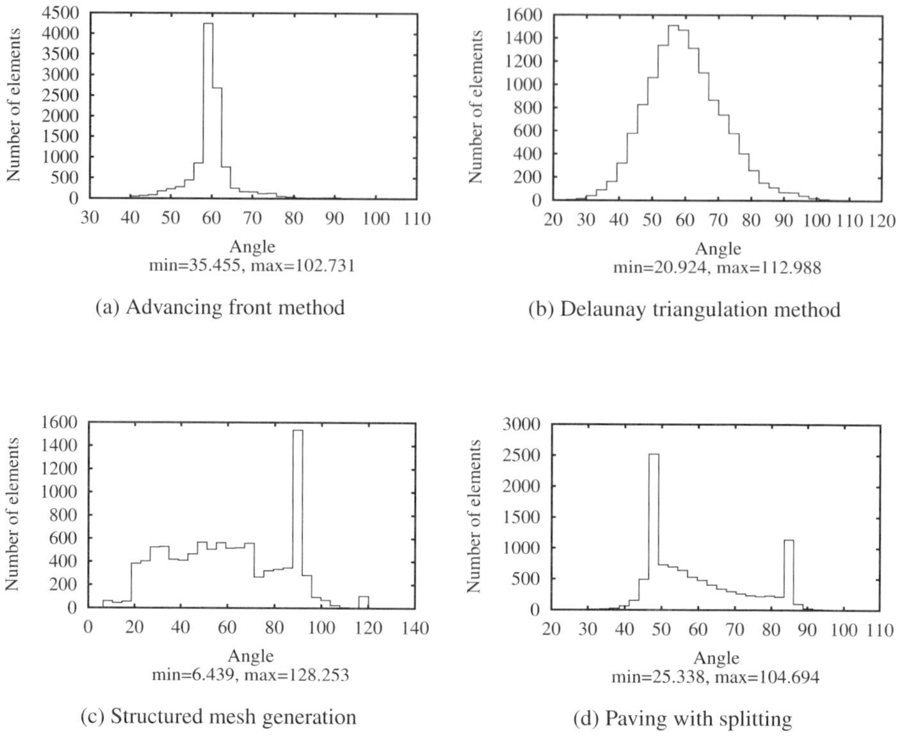

Figure 10.5: Distribution of the internal angles.

the same number of elements except the triangle halving method which generates a few more elements. Figure 10.6 shows the meshes generated and Figures 10.7 to 10.9 show the different shape measures for the meshes. From the figures it can be concluded that the paving method can generate highly regular quadrilateral elements but at the same time it can also generate the worst quality elements. The triangle halving method produces the worst quality elements. The Rank and the fission methods generate good to average quality elements and in addition they do not tend to produce degenerate quadrilateral elements.

10.2.3 The outline of Scotland

The outline of the Scottish mainland was meshed as a demonstration of the power of the Delaunay triangulation method detailed in Chapter 6. Scotland's coastline is well known to be very rugged and the multiple sealochs result in a highly non-convex shape.

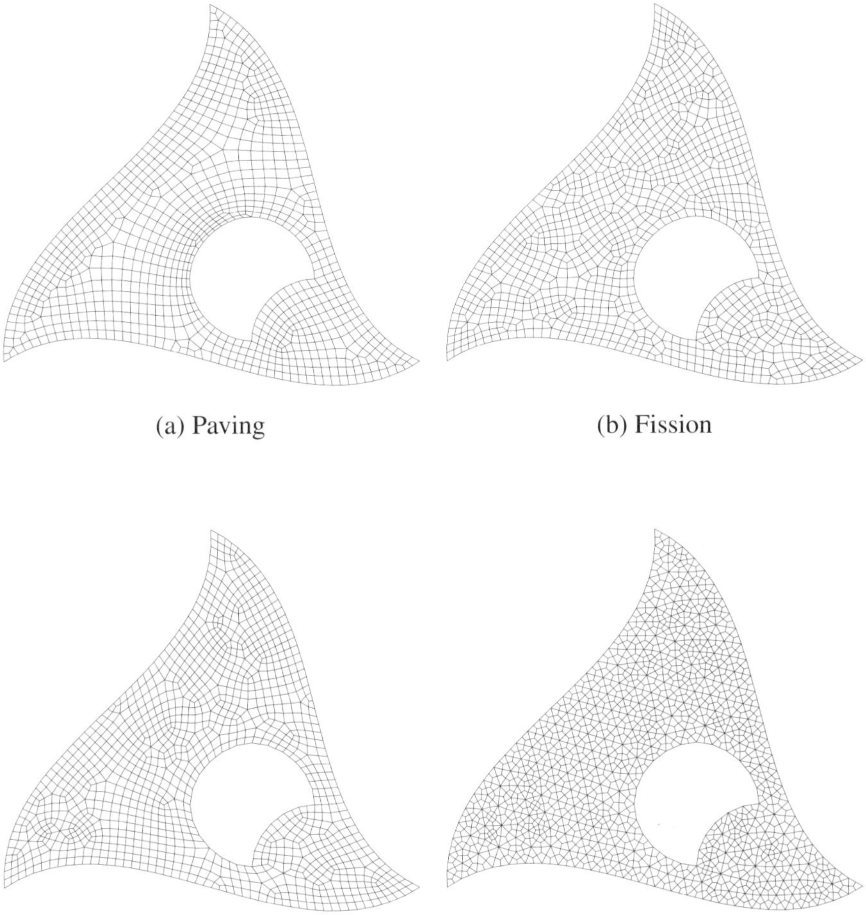

(a) Paving (b) Fission

(c) Rank method (d) Halving

Figure 10.6: Meshes with quadrilateral elements.

The initial triangulation of the course mesh was obtained from an ordered list of boundary coordinates, 635 for the coastline, and 33 for the loch boundaries. These boundaries were triangulated using a custom made program built on the library of mesh generation functions described in this book.

The generated coarse triangulation was remeshed using the Delaunay method with (`CENTROID` variation, $\alpha = 1.5$, $\beta = 2.0$). A local control source file was assembled to create the boundary zone in which refinement is required. The resulting mesh is shown in Figure 10.10. This mesh contains 10,836 triangular elements and 6,243

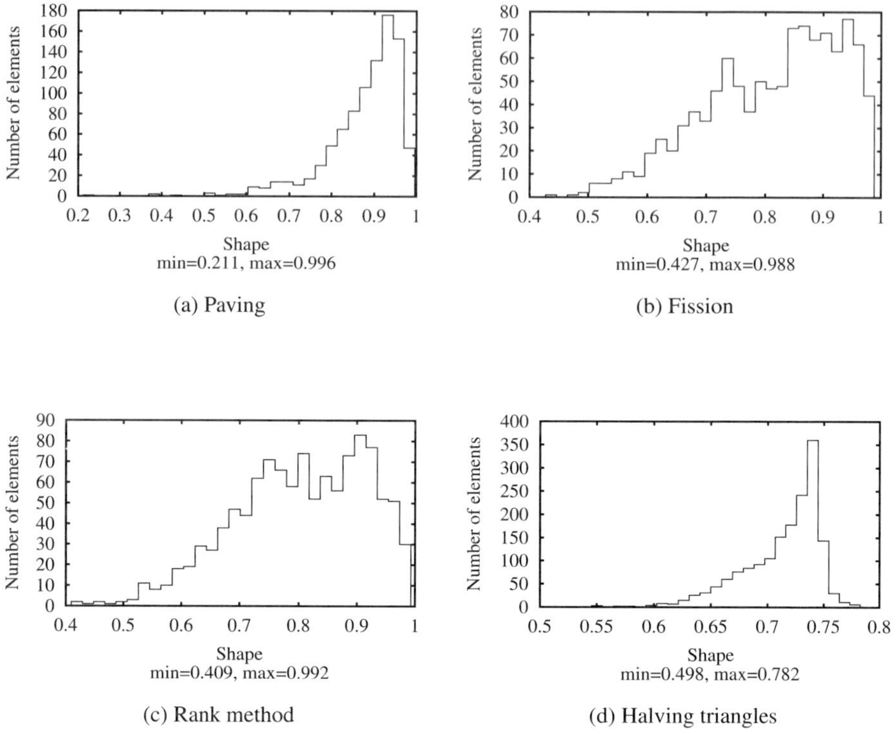

Figure 10.7: Distribution of the shape measure discussed in Section 5.4.2.

nodes. The full generation process, including I/O, pre and postprocessing took less than two minutes (106 seconds) on a single IRIX64 IP30 processor.

10.2.4 An adaptive meshing example

Figure 10.11 shows a uniform coarse mesh of a circuit board, created with the advancing front method. A preliminary analysis of the stress loading of the board was carried out. Error analysis was performed on the stress results based on a simple nodal averaging procedure [2]. Consequently the errors were converted into new mesh parameters for an adaptive remeshing of the board. The resulting non-uniform mesh is shown in Figure 10.12.

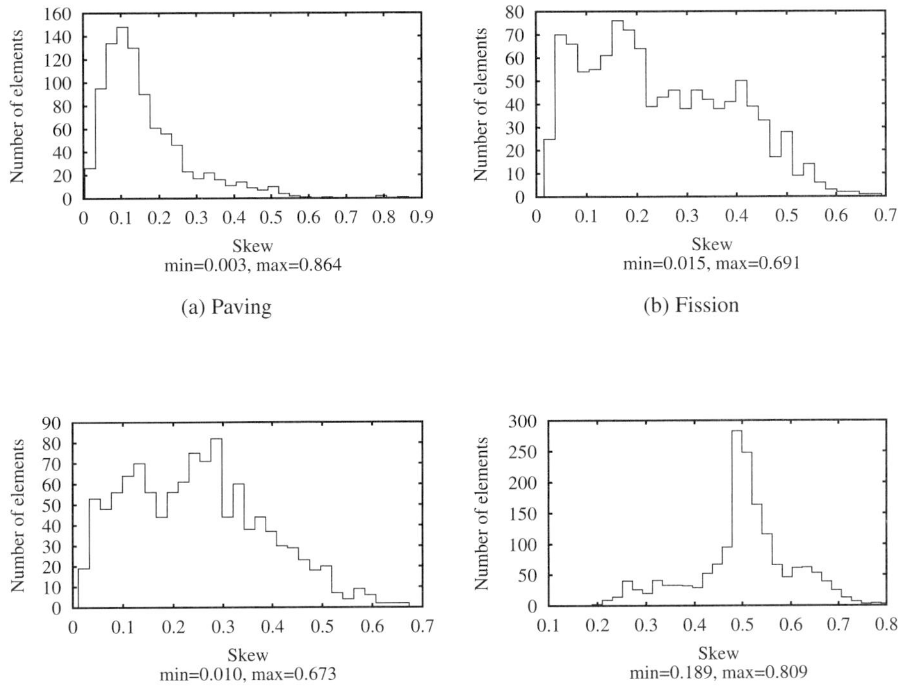

(a) Paving (b) Fission

(c) Rank method (d) Halving triangles

Figure 10.8: Distribution of the skew shape measure defined in section 9.5.

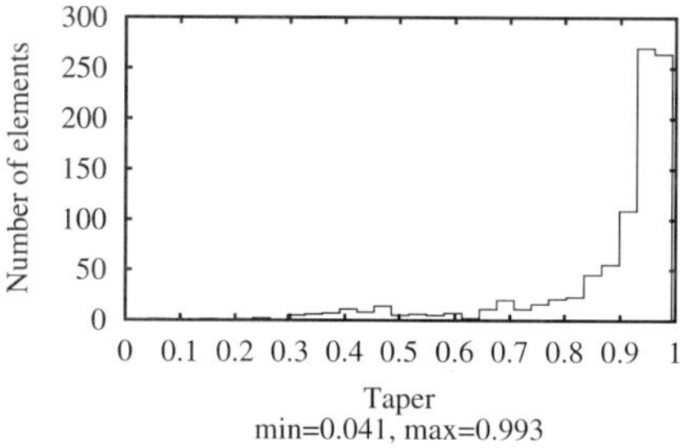

(a) Paving

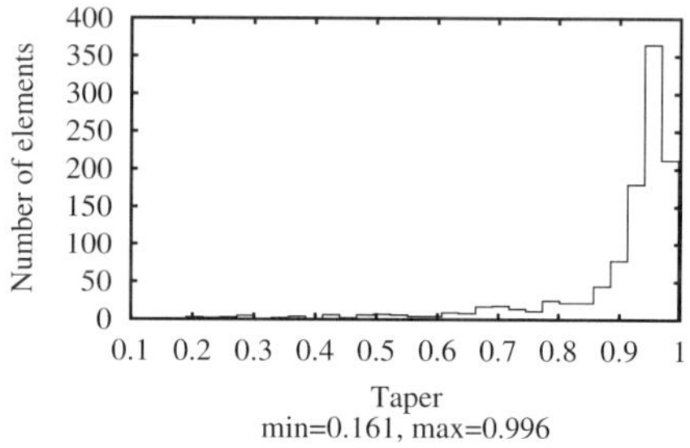

(b) Fission

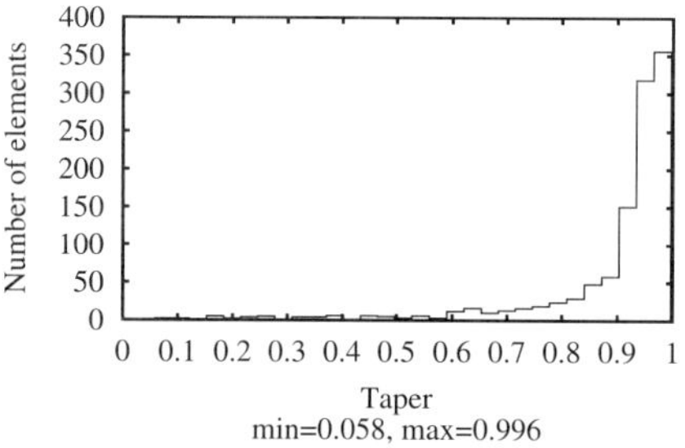

(c) Rank method

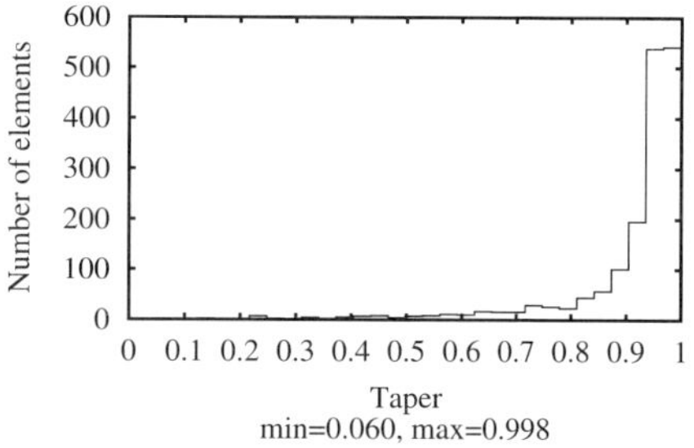

(d) Halving triangles

Figure 10.9: Distribution of the taper shape measure defined in section 9.5.

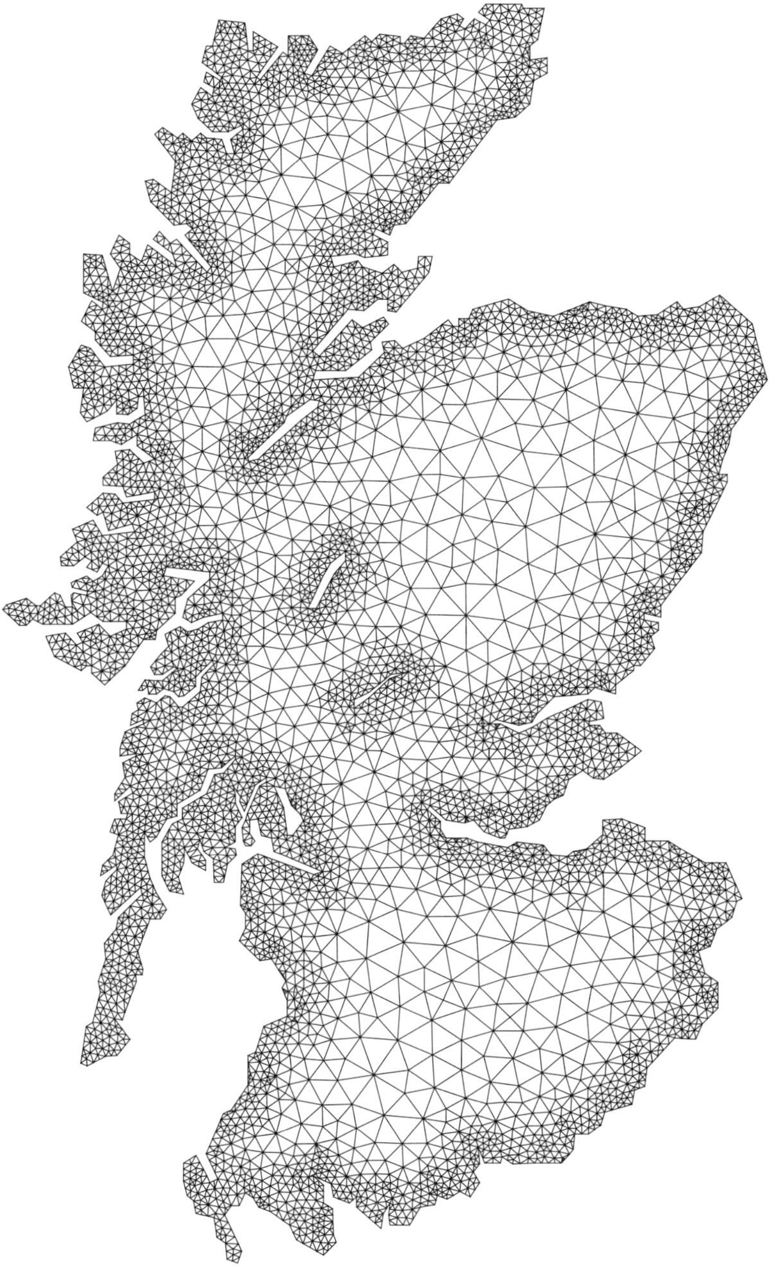

Figure 10.10: The meshed outline of Scotland.

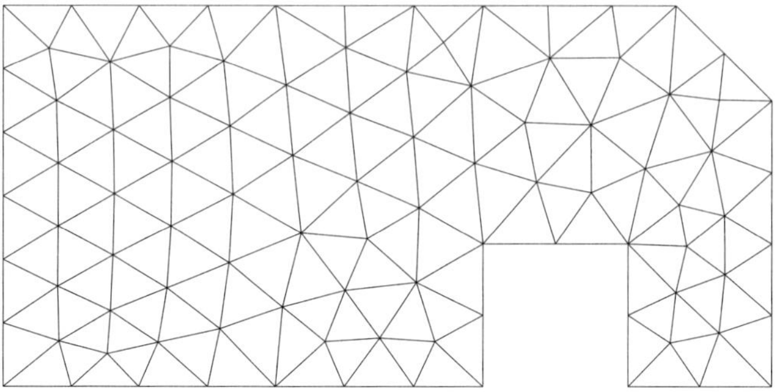

Figure 10.11: Uniform meshing: coarse mesh.

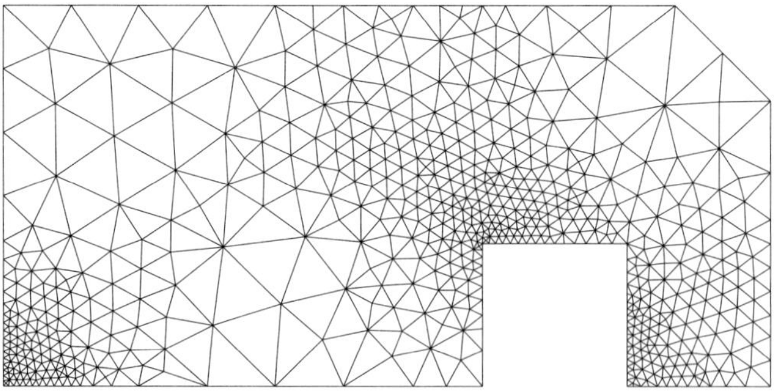

Figure 10.12: Adaptive meshing: adaptive mesh.

Mesh	No. of elements	No. of points	No. of sliver elements	Dihedral angle [degrees]	
				min	max
Tetrah	1344	368	2	15.81	154.79
Cube	10243	2159	6	8.79	157.85
Sphere	22518	4118	8	1.85	176.54
LCube	1191	321	3	9.38	160.90

Table 10.3: Some properties of the 3D tetrahedra meshes.

10.3 Three-dimensional meshes

Figure 10.13 shows four benchmark meshes to be meshed by the three dimensional Delaunay mesh generator program described in Chapter 7. The shape of the first three meshes complies with the benchmark specification described in Reference [1]. The input surface meshes were generated using the 2D Delaunay mesh generator. The properties of the meshes are presented in Table 10.3 and Figures 10.14 to 10.17.

10.3.1 Surface meshing of a knot

A three-dimensional mesh without three-dimensional elements is the following example. A knot constructed out of 12 beams was surface-meshed with the Delaunay method. The result is shown in Figure 10.18.

10.3.2 Volume meshing around a wheel

The wheel object shown in Figure 10.19 was placed inside a cube. The volume around the wheel was consequently remeshed using the three-dimensional Delaunay tetrahedralization method presented in Chapter 7. The top half of the resulting volume mesh was cut off to reveal the presence of the (surface meshed) wheel object inside. A shading of the result is shown in Figure 10.20.

Bibliography

[1] Dompierre, J., Labbé, P., Guibault, F., Camarero, R., "Proposal of Benchmarks for 3D Unstructured Tetrahedral Mesh Optimization", Rapport CERCA R98-91, École Polytechnique Montréal, Centre de Recherche en Calcul Appliqué 5160, boul. Décarie, suite 400, Montréal, (Québec), H3X 2H9, Canada, September 1998, Presented at 7th International Meshing Roundtable, Dearborn, Michigan, 26–28 October 1998.

[2] Topping, B.H.V., Khan, A.I., *Parallel Finite Element Computations*, Saxe-Coburg Publications, Edinburgh, UK, 1996.

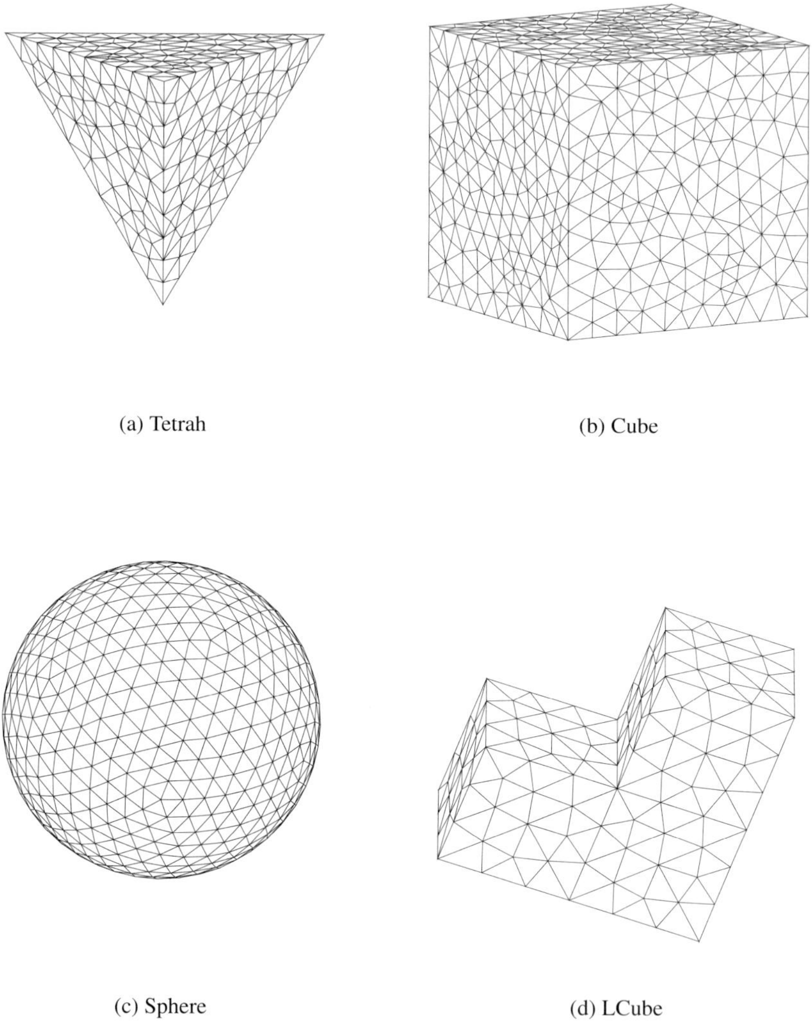

(a) Tetrah (b) Cube (c) Sphere (d) LCube

Figure 10.13: Simple surface benchmark meshes.

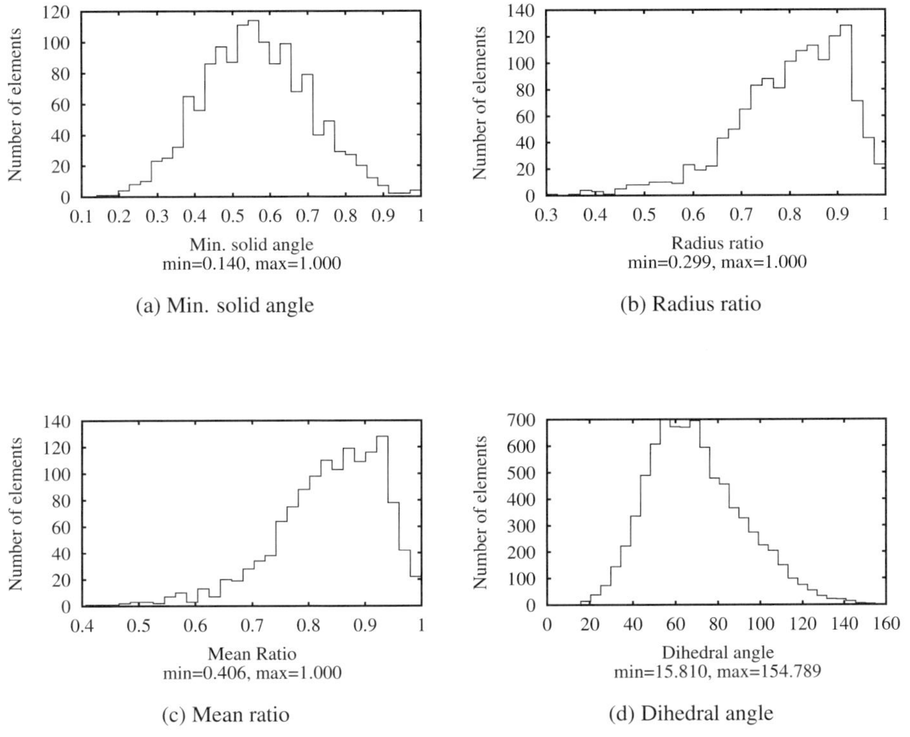

(a) Min. solid angle

(b) Radius ratio

(c) Mean ratio

(d) Dihedral angle

Figure 10.14: Quality measures of the "Tetrah" mesh.

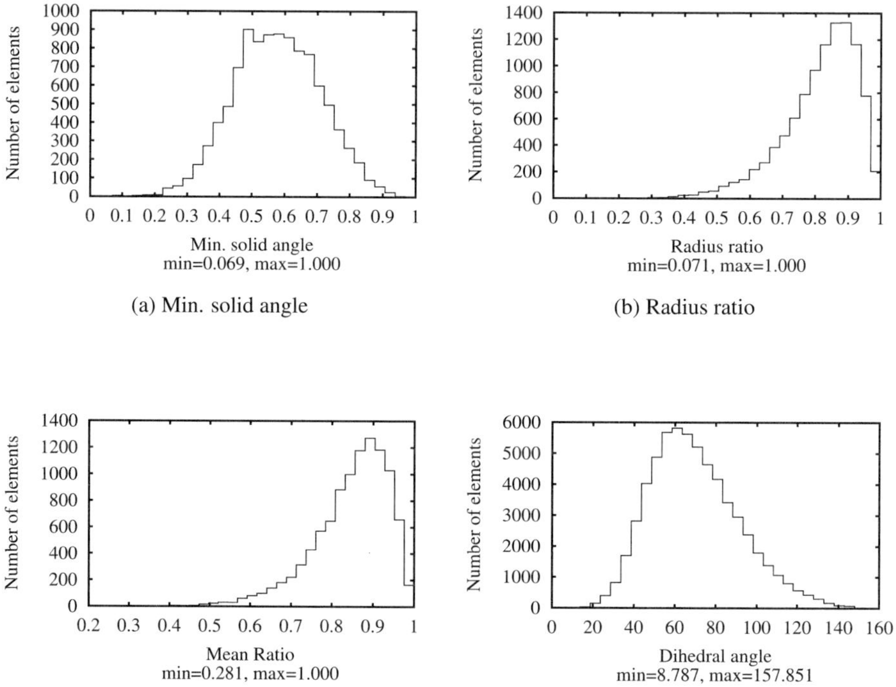

(a) Min. solid angle

(b) Radius ratio

(c) Mean ratio

(d) Dihedral angle

Figure 10.15: Quality measures of the "Cube" mesh.

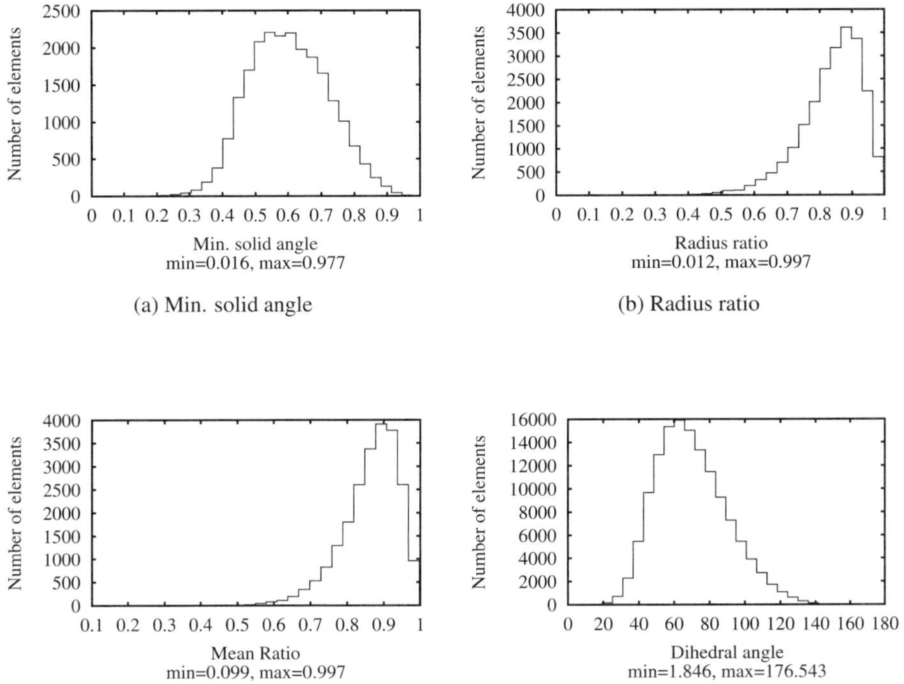

(a) Min. solid angle

(b) Radius ratio

(c) Mean ratio

(d) Dihedral angle

Figure 10.16: Quality measures of the "Sphere" mesh.

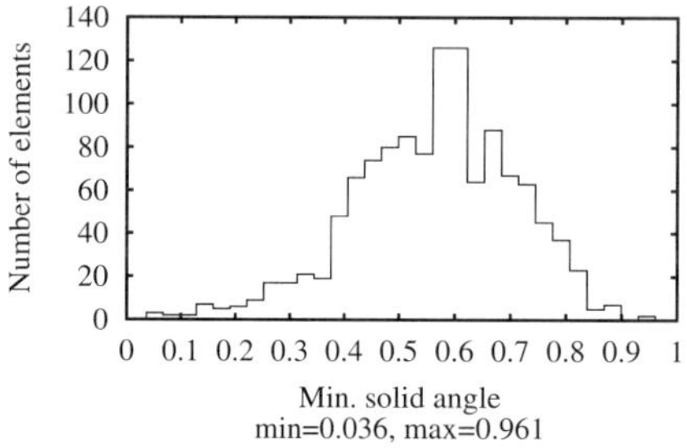

(a) Min. solid angle

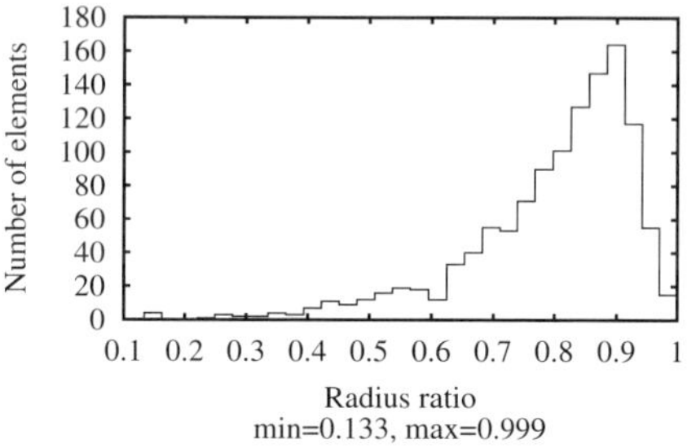

(b) Radius ratio

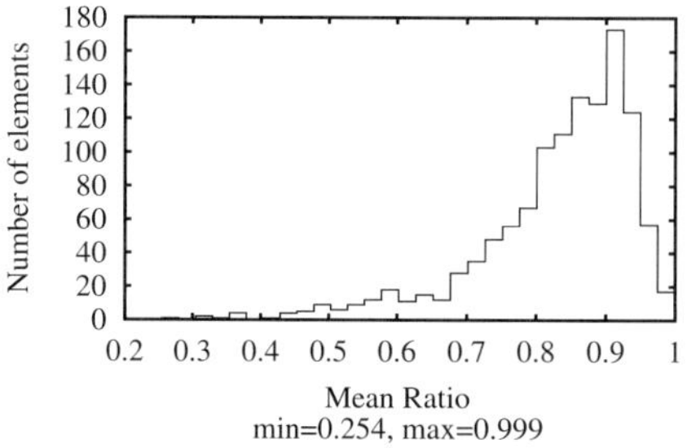

(c) Mean ratio

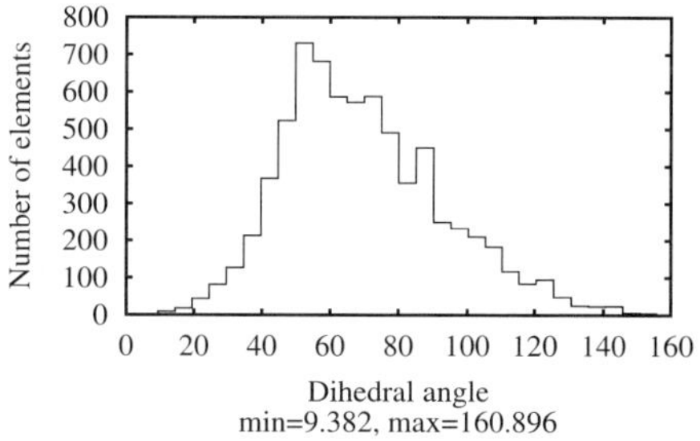

(d) Dihedral angle

Figure 10.17: Quality measures of the "LCube" mesh.

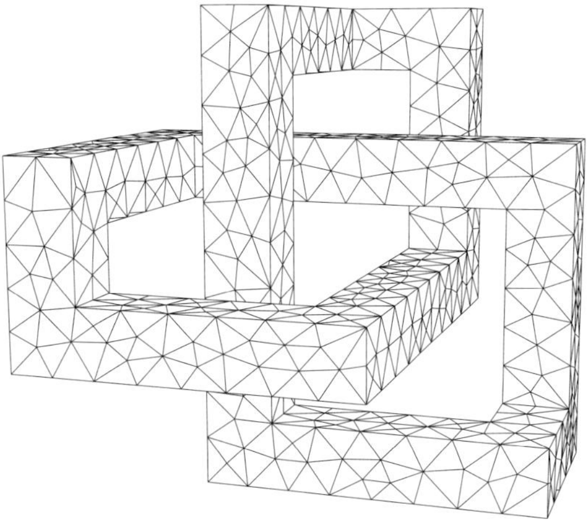

Figure 10.18: Surface meshing of a knot.

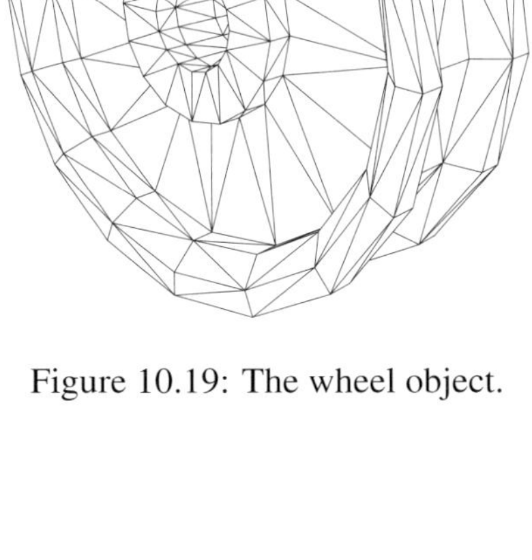

Figure 10.19: The wheel object.

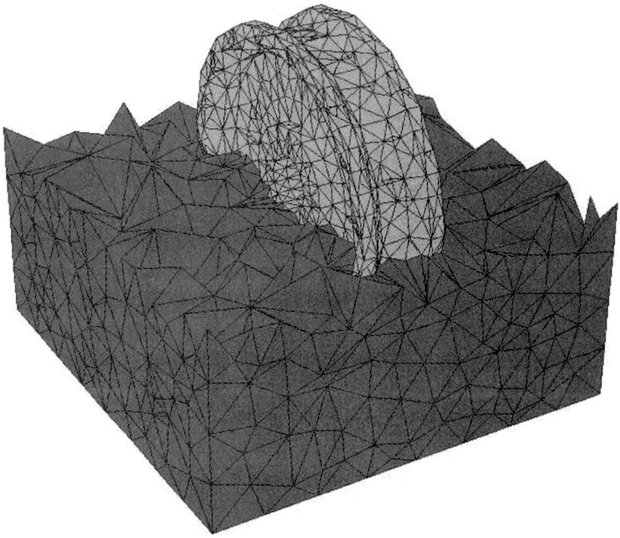

Figure 10.20: Volume meshing around a wheel.

Appendix A

Creating nodes on lines and curves

A.1 Introduction

This appendix explains the algorithms for straight and curve segment refinement. These algorithms are part of the E_Library (described in Appendix D.1) and they are the basic building blocks for all the methods described in this book.

- The `e_SegmentRefine` function can be used to refine a straight line segment according to mesh parameters given in the end nodes.

- The `e_CurveRefine` function can be used to refine a segment represented by a NURBS curve connecting two nodes. The refinement is undertaken using the mesh parameters of the nodes at the end of the curve.

The general rule is that new nodes will be inserted along the segment or curve in such a way that the mesh parameter criteria are met. In the E_Library mesh parameters are defined as lengths assigned to nodes. A segment between two nodes satisfies the mesh parameter criteria if the segment length does not exceed either of the mesh parameter values assigned to its end nodes as shown in Figure A.1(a). If the mesh parameter criteria is not met then additional nodes must be introduced along the length of the segment.

The newly inserted nodes will be assigned a mesh parameter which is derived from the mesh parameters of the end points of the original segment through interpolation. All new segments will satisfy the criteria posed by the new interpolated mesh parameters as shown in Figure A.1(b).

Both functions follow the same algorithmic strategy which will be explained in detail for the straight line function only in Section A.2. Peculiarities relating to the curve refinement will be discussed in greater detail in Section A.3.

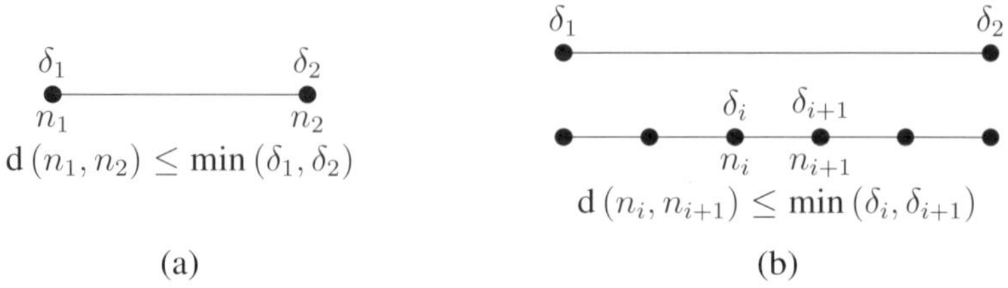

Figure A.1: The definition of mesh parameter criteria. (a) The mesh parameters are satisfied. (b) Nodes are inserted until the mesh parameter criteria are satisfied.

A.2 Refining straight line segments

In this section the `e_SegmentRefine` function for refining straight segments is described.

A.2.1 Syntax

The syntax of the segment refinement function is:

```
int e_SegmentRefine(double    *start_crd,
                    double     start_mpr,
                    double    *end_crd,
                    double     end_mpr,
                    double  ***seg_crd,
                    double   **seg_mpr,
                    int        generate_nodes,
                    int        mode)
```

A.2.2 Function description

This function generates nodes on the straight line between two points in space according to mesh parameters given for those points. The line starts at the point with coordinates:

```
(start_crd[0], start_crd[1], start_crd[2])
```

The line ends at the point given by coordinates:

```
(end_crd[0], end_crd[1], end_crd[2])
```

The mesh parameter at the start of the line is:

```
start_mpr
```

The mesh parameter at the end of the line is:

```
end_mpr
```

If `generate_nodes` has the value `TRUE`, nodes will be generated. The coordinates of the interpolated nodes will be put in an array of which the address is given by:

```
seg_crd
```

The values of the interpolated mesh parameter will be put in a vector of which the address is given by:

```
seg_mpr
```

Both array and vector space will be allocated by this function. Any information already at these addresses will be lost.

If `generate_nodes` has the value `FALSE` no nodes will be generated. The value of `seg_crd` and `seg_mpr` will be discarded. The value of `mode` can be any of the following three pre-defined macros:

- `E_SEGREF_NONE` if both even and odd numbers of generated nodes are acceptable
- `E_SEGREF_ODD` if an odd number of nodes must be generated
- `E_SEGREF_EVEN` if an even number of nodes must be generated

`mode` is not taken into account when no nodes are generated because the mesh parameters are too big.

The return value of the function is the number of nodes that is or would be generated. The interpolated nodes are generated in the order from start to end.

A.2.3 Algorithmic implementation

The relation between the length of the unrefined segment and the values of the mesh parameters determines which implementation is followed. It is trivial when (according to Figure A.1(a) both δ_1 and δ_2 are larger than the segment length, no new nodes have to be created as the original segment already satisfies the mesh parameter criteria.

Different procedures are used depending on whether one or both the mesh parameters are smaller than the original segment length.

A.2.3.1 Both mesh parameters are smaller than the original segment length

Assume for a moment that the mesh parameter at the start of the segment is smaller than the mesh parameter at the end of the segment. Referring to Figure A.2 the lengths of the new segments created by the node insertion are ideally equal to the smaller of the two mesh parameters at the ends of the new segment.

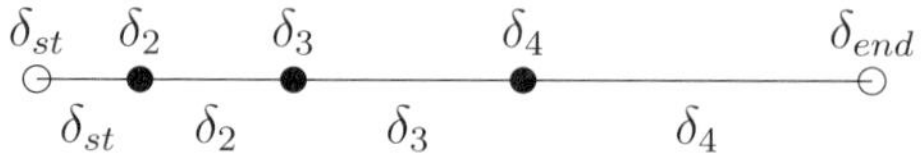

Figure A.2: The ideal lengths of the new segments.

The total length of the segment s (obtained as the cartesian distance between the start and the end points of the segment) can therefore also be calculated as the sum of the ideal lengths of the new segments:

$$s = \delta_{st} + \delta_2 + \delta_3 + \delta_4 \tag{A.1}$$

However for a smooth transition from that start mesh parameter δ_{st} to the end mesh parameter δ_{end} the new intermediate mesh parameters should be resolved using interpolation. A power series with coefficient α is used.

$$\begin{aligned} s &= \delta_{st} + \alpha\delta_{st} + \alpha^2\delta_{st} + \alpha^3\delta_{st} && \text{(A.2)} \\ &= \delta_{st}\left(1 + \alpha + \alpha^2 + \alpha^3\right) && \text{(A.3)} \\ &= \frac{\delta_{st}\left(1 - \alpha^{n+1}\right)}{(1 - \alpha)} && \text{(A.4)} \end{aligned}$$

where n is the number of new nodes that have to be inserted.

Taking into account logical continuation of the series, the mesh parameter at the end of the original segment can be written as:

$$\delta_{end} = \alpha^{n+1}\delta_{st} \tag{A.5}$$

From this information both α and the number of new nodes n can be calculated:

$$\begin{aligned} s &= \frac{\delta_{st}\left(1 - \frac{\delta_{end}}{\delta_{st}}\right)}{(1 - \alpha)} && \text{(A.6)} \\ s &= \frac{\delta_{st} - \delta_{end}}{(1 - \alpha} && \text{(A.7)} \\ s - s\alpha &= \delta_{st} - \delta_{end} && \text{(A.8)} \\ \alpha &= \frac{s - \delta_{st} + \delta_{end}}{s} && \text{(A.9)} \end{aligned}$$

and

$$\alpha^{n+1} = \frac{\delta_{end}}{\delta_{st}} \tag{A.10}$$

$$n = \frac{\ln \frac{\delta_{end}}{\delta_{st}}}{\ln \alpha} - 1 \tag{A.11}$$

As a result of the nature of the calculation of n it is extremely unlikely that n will be an integer. Therefore the right (integer) number of new nodes $\bar{n}$ can be chosen as the smallest integer larger than n.

$$\bar{n} = \lceil n \rceil \tag{A.12}$$

At this stage the value of variable `mode` can be taken into account. If `mode` equals `E_SEGREF_ODD` it should be verified that $\bar{n}$ is an odd integer and if not increased by 1. When `mode` equals `E_SEGREF_EVEN` and $\bar{n}$ is an odd number then the value of $\bar{n}$ has to be increased to become an even integer. If `mode` equals `E_SEGREF_NONE` no checking has to be performed. Corresponding with $\bar{n}$ a new ratio $\bar{\alpha}$ can be determined using

$$\bar{\alpha} = \left(\frac{\delta_{end}}{\delta_{st}} \right)^{\frac{1}{\bar{n}+1}} \tag{A.13}$$

If the total length of the original segment is calculated again using Equation (A.4) a new length $\bar{s}$ is obtained

$$\bar{s} = \frac{\delta_{st}\left(1 - \bar{\alpha}^{\bar{n}+1}\right)}{(1 - \bar{\alpha})} \tag{A.14}$$

The increase of n to $\bar{n}$ causes $\bar{s} \geq s$. Therefore the new total length $\bar{s}$ is "squeezed back" to the original total length s using a scaling factor β as shown in Figure A.3

$$\beta = \frac{s}{\bar{s}} \tag{A.15}$$

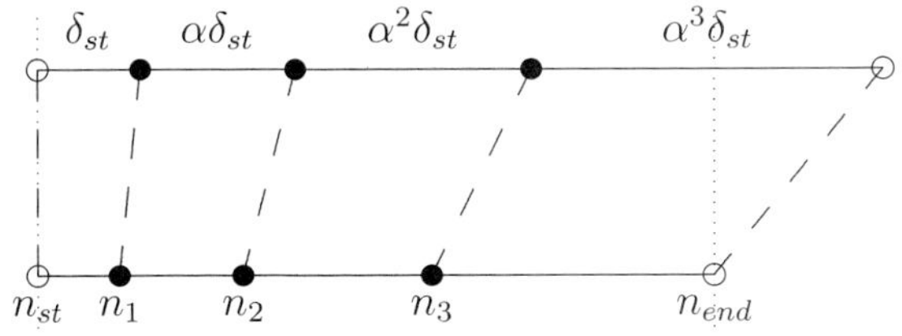

Figure A.3: "Squeezing" the theoretical point coordinates back to the segment using the scaling factor β.

The location of each new point can therefore be calculated as:

$$\vec{n}_0 = \vec{n}_{st} \tag{A.16}$$

$$\vec{n}_i = \vec{n}_{i-1} + \beta\alpha^{i-1}\delta_{st}\frac{\vec{n}_{end} - \vec{n}_{st}}{s} \tag{A.17}$$

And the corresponding interpolated mesh parameter values will be

$$\delta_0 = \delta_{st} \tag{A.18}$$

$$\delta_i = \delta_{i-1} + \beta\alpha^{i-1}\delta_{st}\frac{\delta_{end} - \delta_{st}}{s} \tag{A.19}$$

When δ_{st} equals δ_{end} (or the difference between the two values is smaller than a threshold value) a major simplification can be made to the entire process because all new segments will be of equal length (ideally $\delta = \delta_{st} = \delta_{end}$). The number of new nodes to be inserted is easily found as:

$$\bar{n} = \left\lceil \frac{s}{\delta} \right\rceil - 1 \tag{A.20}$$

The value of $\bar{n}$ might have to be increased according to the requirements of `mode`. The new coordinates and mesh parameters are found by:

$$\vec{n}_i = \vec{n}_{st} + \frac{i}{\bar{n}+1}\left(\vec{n}_{end} - \vec{n}_{st}\right) \tag{A.21}$$

$$\delta_i = \delta \tag{A.22}$$

A.2.3.2 Only one mesh parameter is smaller than the original segment length

When only one of the mesh parameters is smaller than the length of the original segment a different strategy has to be adopted. The original segment does not satisfy the mesh criteria in the node which has the smaller mesh parameter.

The philosophy is that new nodes will be inserted from the end with the smaller mesh parameter to the end with the larger mesh parameter until no further nodes are required. As shown in Figure A.4 the distance that each new node is positioned away from the previous node equals the mesh parameter value at the previous node. The interpolation of the mesh parameter values for the new nodes is linear.

Assuming for that δ_{st} is smaller than δ_{end} and that therefore $\delta_{st} < s < \delta_{end}$ then the ideal position for a new node n_i^* can be calculated with the following recursive pattern of equations:

$$\delta_0^* = \delta_{st} \tag{A.23}$$

$$\vec{n}_0^* = \vec{n}_{st} \tag{A.24}$$

$$\vec{n}_i^* = \vec{n}_{i-1}^* + \frac{\delta_{i-1}^*}{s}\left(\vec{n}_{end} - \vec{n}_{st}\right) \tag{A.25}$$

$$\delta_i^* = \delta_{i-1}^* + \frac{\delta_{i-1}^*}{s}\left(\delta_{end} - \delta_{st}\right) \tag{A.26}$$

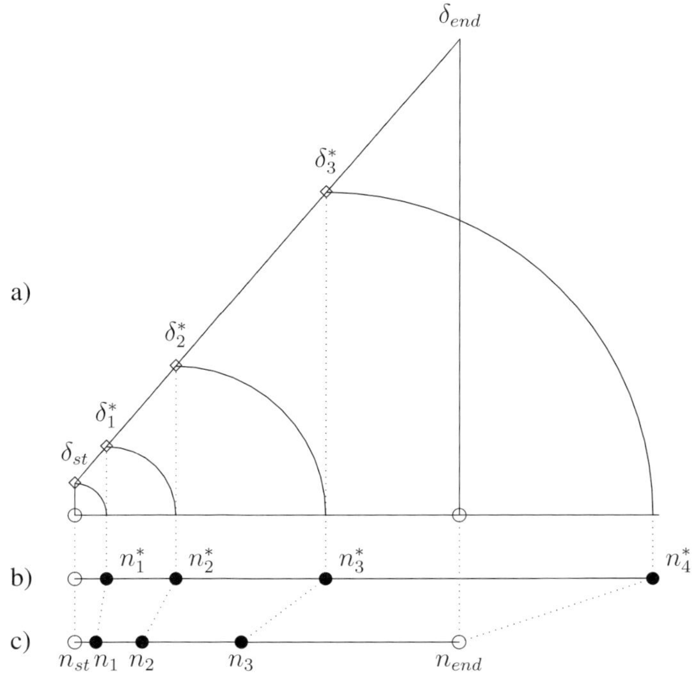

Figure A.4: Creation of new nodes when only one mesh parameter is smaller than the segment length.

The result of the calculation of the ideal positions is shown in Figure A.4(b). Only exceptionally will the last node coincide with the end node of the segment. An overshoot is more likely. By analogy with the procedure described in the previous section the total length between the starting node and the last new node is "squeezed back" to the original length of the segment using a scale parameter β.

$$\bar{s} = \|\vec{n}_n^* - \vec{n}_{st}\| \tag{A.27}$$

$$\beta = \frac{s}{\bar{s}} \tag{A.28}$$

The final positions of the new nodes are therefore:

$$\delta_0 = \delta_{st} \tag{A.29}$$

$$\vec{n}_0 = \vec{n}_{st} \tag{A.30}$$

$$\vec{n}_i = \vec{n}_{i-1} + \beta \frac{\delta_{i-1}}{s} (\vec{n}_{end} - \vec{n}_{st}) \tag{A.31}$$

$$\delta_i = \delta_{i-1} + \beta \frac{\delta_{i-1}}{s} (\delta_{end} - \delta_{st}) \tag{A.32}$$

If the value of `mode` has to be taken into account an extra node might have to be "squeezed in". Equations (A.29) to (A.32) remain unaltered even though two nodes will fall outside the original segment.

A.2.4 Examples

In the examples shown in Table A.1, the distance (s) between start and end nodes is conveniently set to 100. The numbers above the new nodes are the interpolated mesh parameters. The numbers below the new segments are the segment lengths. The chosen `mode` is shown in the column `m` where `N` stands for `E_SEGREF_NONE`, `O` for `E_SEGREF_ODD` and `E` for `E_SEGREF_EVEN`.

δ_{st}	δ_{end}	m	result of `e_SegmentRefine`
20	20	N	20 20 20 20 20 20 20 20 20
20	20	O	20 20 20 20 20 16.7 16.7 16.7 16.7 16.7 16.7
15	15	N	15 15 15 15 15 15 14.3 14.3 14.3 14.3 14.3 14.3 14.3
10	30	N	11.5 13.4 15.6 18.3 21.5 25.4 7.7 9.3 11.1 13.4 16.1 22.3 23.2
20	120	N	34.3 62.9 14.3 28.3 57.1
20	120	O	26.7 40 66.7 6.7 13.3 26.7 53.3

Table A.1: Examples of `e_SegmentRefine`.

A.3 Refining curved segments

In this section the e_CurveRefine function for refining curve segments is described.

A.3.1 Syntax

The syntax of the curve segment refinement function is:

```
int e_CurveRefine(NURBSCURVE *curve,
                  double    start_t,
                  double    start_mpr,
                  double    end_t,
                  double    end_mpr,
                  double ***seg_crd,
                  double  **seg_mpr,
                  double  **seg_t,
                  int       generate_nodes,
                  int       mode)
```

A.3.2 Function description

This function generates nodes on the NURBS curve segment defined by curve and delimited by two curve parameters. The refinement is carried out according to mesh parameters assigned to the delimiting points.

The curve segment starts at the point with the curve parameter the start_t and ends at the point with the curve parameter end_t.

The mesh parameter at the start of the curve segment is:

```
start_mpr
```

The mesh parameter at the end of the curve segment is:

```
end_mpr
```

If generate_nodes has the value TRUE, nodes will be generated. The coordinates of these interpolated nodes will be put in an array of which the address is given as:

```
seg_crd
```

The values of the interpolated mesh parameter will be put in a vector of which the address is given as:

```
seg_mpr
```

The values of the interpolated curve parameter will be put in a vector of which the address is given as:

```
seg_t
```

Both array and vector space will be allocated by this function. Any information already on these addresses will be lost.

It is important to note that in contrast to `e_SegmentRefine` the coordinates at start and end points of the curve segment will also be included in `seg_crd` and in the return value.

If `generate_nodes` has the value `FALSE`, no nodes will be generated. The value of `seg_crd` and `seg_mpr` will be discarded.

The value of `mode` can be any of the following three pre-defined macros:

- `E_SEGREF_NONE` if both even and odd numbers of generated nodes are acceptable
- `E_SEGREF_ODD` if an odd number of nodes must be generated
- `E_SEGREF_EVEN` if an even number of nodes must be generated

`mode` is not taken into account when no nodes are generated when the mesh parameters are too big.

The return value of the function is the number of nodes that is or would be generated. The interpolated nodes are generated in the order from start to end.

A.3.3 Algorithmic implementation

The implementation of `e_CurveRefine` is identical to that of `e_SegmentRefine`, the only difference being the calculation of lengths. Whereas with a straight line the lengths of segments can be calculated as ordinary cartesian distances, the lengths of curved segments must be obtained in a different way.

Three curve related functions are at the programmer's disposal. (See the E_Library header file for correct usage of these functions).

- `crv_getlength` which determines the length of a curved segment given by a NURBS curve definition structure and delimited by two curve parameter values according to a specified accuracy level.
- `crv_getcoords` which calculates the coordinates of a point on a curved segment given by a NURBS curve definition structure and specified by a given curve parameter value.

- `crv_length2param` which determines the curve parameter of a point on a given curved segment that is a specified length away from a previous point determined by a single curve parameter. This function will for example determine what curve parameter the point has which is 10 length units away from the point with curve parameter 0.3 on a curve between parameters 0.0 and 1.0 of total length 100. Additional required inputs for this function are the direction in which the specified length should be travel and the accuracy level by which all lengths should be determined.

A.3.4 Example

The following example demonstrates the capacities of `e_curve_refine`. The example curve is defined by the following parameters:

```
crv->
  Degree      3       # quadratic curve
  EndPoint1 {0.0,0.0,0.0}
                      # point corresponding with t=0.0
  EndPoint2 {10.0,0.0,0.0}
                      # point corresponding with t=1.0
  CntlPoints[0] {3.0,3.0,0.0}
                      # first control point
  CntlPoints[1] {7.0,-3.0,0.0}
                      # second control point
  Weights[0] 1.0      # weight at t=0.0
  Weights[1] 0.8      # weight at first control point
  Weights[2] 0.8      # weight at second control point
  Weights[3] 1.0      # weight at t=1.0
```

The example call to the curve refinement function is:

```
n = e_CurveRefine(
    crv,              # NURBSCURVE definition
    0.0,              # crv. param. at start of crv. segment
    0.2,              # mesh param. at start of crv. segment
    1.0,              # crv. param. at end of crv. segment
    2.0,              # mesh param. at end of crv. segment
    seg_crd,
    seg_mpr,
    seg_t,
    TRUE,               # generate nodes
    E_SEGREF_ODD)       # mode
```

With this data 15 new internal segment nodes were created ($n = 17$). The numerical values are tabulated in Table A.2. Their location is shown in Figure A.5.

	coordinates		curve	mesh	segment
n	x	y	param	param	length
1	0.0	0.0	0.0	0.2	
2	0.12	0.11	0.02	0.23	0.17
3	0.27	0.24	0.04	0.26	0.19
4	0.45	0.37	0.06	0.30	0.23
5	0.68	0.50	0.09	0.34	0.26
6	0.96	0.62	0.12	0.39	0.31
7	1.30	0.71	0.16	0.46	0.36
8	1.72	0.77	0.20	0.53	0.42
9	2.20	0.76	0.25	0.61	0.48
10	2.76	0.69	0.30	0.70	0.57
11	3.40	0.54	0.36	0.81	0.66
12	4.14	0.31	0.43	0.95	0.77
13	4.98	0.01	0.50	1.10	0.90
14	5.97	-0.34	0.58	1.27	1.04
15	7.14	-0.67	0.69	1.48	1.22
16	8.55	-0.74	0.83	1.72	1.42
17	10.0	0.0	1.0	2.0	1.66

Table A.2: Example numerical values.

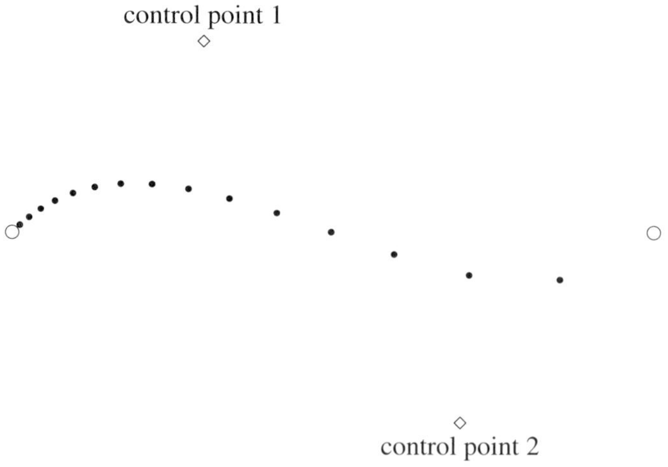

Figure A.5: Example results of `e_curve_refine`.

Appendix B

Block structure definition file format

B.1 Introduction

In this appendix the file format of the structured mesh generation program is described. The file format is divided into several parts: header, entity definition, boundary condition definition. In this context an entity is a one or two dimensional element, for example a surface patch. The appendix also tries to provide a detailed explanation for the different aspects of the file format through examples.

B.2 Entity definitions

Figure B.1 shows the conventions used for the definition of the different entity types. The origin of the local or element coordinate system is the first node of the geometric entity. The 2D surfaces are defined in an anti-clockwise manner which ensures that the final finite elements will also be defined in an anti-clockwise manner. The first local coordinate direction (ξ) is parallel to the edge defined by using the direction defined by the first and the second nodes of the geometric entity. In the case of 2D geometric entities the second local coordinate direction (η) is parallel to the edge determined using the direction defined by the first and the last node of the entity. The first and second nodes determine the first side, the second and third nodes specify the second side and so on. The nodes are created and numbered in a row as is also shown in Figure B.1 for all 1D and 2D entity types.

B.3 Data file format

The input to the program is a textual data file which describes the geometric model of the structure. The data structure has been designed in such a way that the identified sequence of mesh definition can be carried out easily at one step. The file starts with

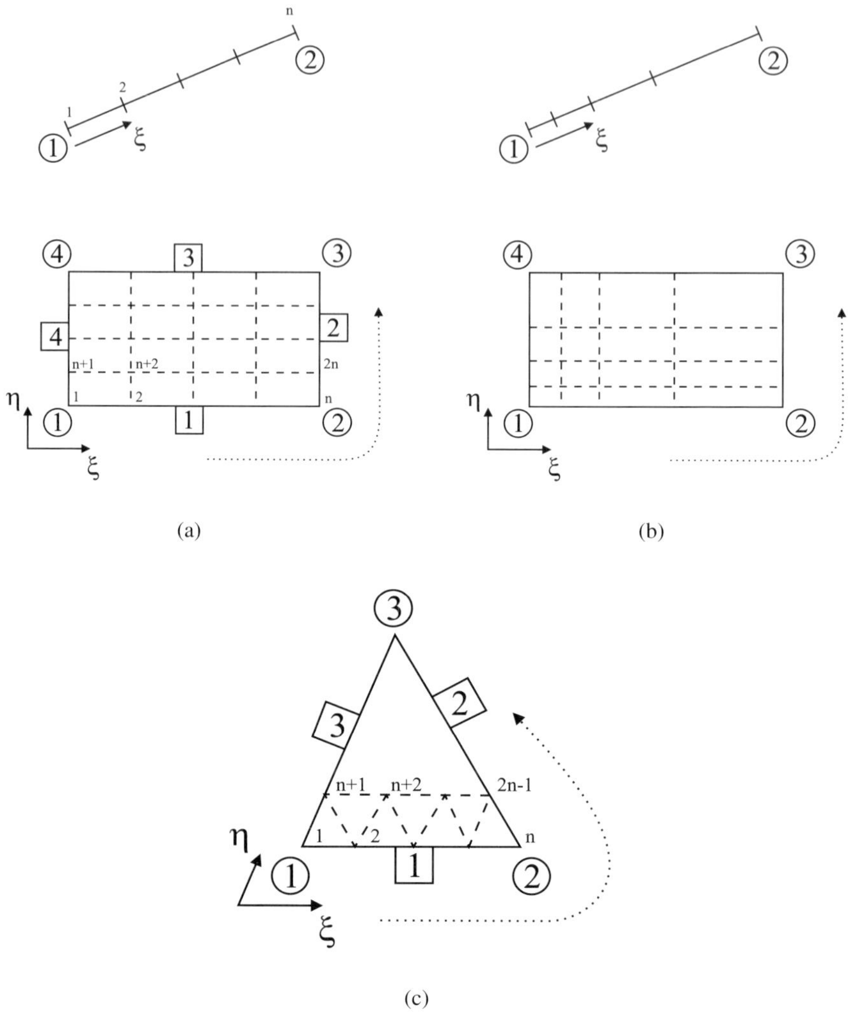

Figure B.1: Conventions for the entity definition, (a) uniform quadrilateral, (b) nonuniform quadrilateral and (c) triangular entities.

a header section, then lists the point entities, the two and the one dimensional entities and finally the boundary conditions. Comment and empty lines can be inserted at any line in the file, because these lines are automatically ignored during the reading of the input file. The comment lines start with a "#" sign. Most of the data is provided in a keyword-data format. The information should be placed into one line, new lines cannot be started in the middle of the data definition.

B.3.1 Header section

```
TITLE "this is the descriptive title"
NNODES          i
NSURFACES       j
NLINKS          k
( GEN_MATERIAL   )
( NMATERIALS   l )
( NBC_NODES    m )
( NBC_SIDES    n )
```

Table B.1: Header section of the geometric data file.

The header section of the input file is defined in Table B.1 where the TITLE keyword defines the title of the project, the NNODES keyword defines the number of point entities, the NSURFACES keyword defines the number of 2D entities and the NLINKS keyword defines the number of 1D entities. Later in the data file the above given number of entities should be specified. The entities are not defined in the order of the number of dimensions, such as first points, then one dimensional and finally two dimensional elements. This is because the two dimensional entities are considered to be the main components of the structure, while the one dimensional entities are considered as additional components. In spite of this, it is possible to generate a structure containing only one dimensional elements.

The bracketed keywords are optional. The GEN_MATERIAL keyword simply specifies that the mesh material information should be generated. The next NMATERIALS keyword provides the maximum number of user definable material types. If only the first keyword is specified, then all finite elements will have the default material definition which is assigned to them according to their type. In this case all of the same element types will have the same material. If the second keyword is defined, then additional material definition may be defined by the user. This means that it is possible to assign different properties from default material to all finite elements generated using the same geometric entity. The geodesic strings [3] (link type 5 according to the E_Library specification [2] or see Section D.1) are exceptions, because their material cannot be overdefined, they will always have the default material definition. The last two keywords NBC_NODES and NBC_SIDES define the number of boundary condition points and the number of boundary condition sides.

B.3.2 Point entity definition

In this section points are defined by their three coordinate values. These points may be used in the later sections to define higher dimension entities. The number of points given in this section must be the same as NNODES defined in the header section, but it is not necessary to use all the defined points. Those points which are not referenced by their index in the later sections are ignored and they will not appear in the final mesh.

The format of this section is shown in Table B.2, where `x1-coord`, `y1-coord` and `z1-coord` provide the position of the first point in the global coordinate system. The definition of all points is similar.

```
NODE_COORDINATES
1  x1-coord  y1-coord  z1-coord
2  x2-coord  y2-coord  z2-coord
.
.
.
i  xn-coord  yn-coord  zn-coord
```

Table B.2: Point entity definition in the geometric data file.

B.3.3 Surface entity definition

In this section the two dimensional entities, the quadrilateral and the triangular surfaces are defined. They can be defined in any order, but their numbering should be consecutive starting from one and the total number of these entities should be equal to `NSURFACES` specified in the header section. At the moment the triangular surfaces can be defined only with straight edges, while the quadrilateral surfaces are defined as cubic patches. A special case of the cubic patch definition is when the quadrilateral surface has straight edges. To help the insertion of one dimensional elements into the generated mesh (such as geodesic strings [3]) those one dimensional elements which are part of the surface are also defined in this section. The format of the surface entity definition is presented in Table B.3.

The bracketed sections are optional, the square brackets ('`[`','`]`') surround options that can be defined only for quadrilateral geometric entities and the brackets ('<', '>') contain expressions only for the triangular geometric entities. The vertical line ('`|`') is written between options, from which at least one should be defined. `n` represents an integer value. The `SHAPE` keyword specifies the form of the surface which can be quadrilateral or triangular shape. A triangular geometric entity is simply defined by its corner nodes after the `DEFINITION` keyword. On the other hand the quadrilateral geometric entity can be defined in three different ways. In the case of the `LINEAR` definition the four corner nodes should be supplied, in the case of the `POINT` definition sixteen nodes should be specified to define a cubic patch and in the case of the `PARAMETRIC` definition four corner nodes and the derivatives are required to define the cubic patch. The order of the derivatives for the parametric definition is:

$$X_\xi \; Y_\xi \; Z_\xi \quad X_\eta \; Y_\eta \; Z_\eta \quad X_{\xi\eta} \; Y_{\xi\eta} \; Z_{\xi\eta} \tag{B.1}$$

```
SURFACE n
  SHAPE          TRIANG | QUAD | 3 | 4
  ELEMENT_TYPE   TRIANG1 | TRIANG3 | TRIANG4 |
                 TRIANG5 | [ QUAD1 ]
  DEFINITION     [ LINEAR | PARAMETRIC | POINT ]
  nodes | [ nodes consts ]
  N_DIVISIONS n [ n ]
  ( DIVISION_WEIGHTS
      weights
      weights
      < weights > )
  ( LINK_DIRECTION   1 | 0   1 | 0   < 1 | 0 >
    LINK_DIVISION
      division
      division
      < division >
    LINK_TYPE
      types
      types
      < types > )
  ( MATERIAL n )
  [ DIAGONAL 1 | 2 | SHORTEST | LONGEST ]
END
```

Table B.3: Surface entity definition in the geometric data file.

A `node` can be a reference to a previously defined point entity or to a generated node of one of the entities. Allowing the latter case of point referencing the definition of the surface topology can be simplified. Two examples are shown in Figure B.2 where the surfaces do not have to be cut up into further pieces to ensure the compatibility between the adjacent surfaces. The format of the *reference point* is either `s n x y` if it references a node in a two dimensional entity or `l n x` if it references a node in a one dimensional entity, where `n` is the index of the entity and `x` and `y` define the position of the node in the local coordinate system of the referenced geometric entity (see Figure B.1). An entity can only reference such an entity which has been defined before the current entity. For example, in surface entity nine `s 1 1 3` references a node in the first surface entity which is in the first column and in the third row of the points to be generated. The example is shown in a quadrilateral and in a triangular surface in Figure B.3.

The `ELEMENT_TYPE` keyword specifies the type of the finite element which will be generated in the geometric entity. Quadrilateral elements cannot be generated in a triangular geometric entity. The name of the element types are references to the available element types in the E_Library specification (see Section D.1) because the generated final finite element mesh will comply the E_Library specification. The `DIVISION` keyword specifies the number of divisions on the sides of the entities. In the case of a triangular entity all sides are divided into an equal number of divisions, but if required the divisions can have different lengths on all sides. In the case of

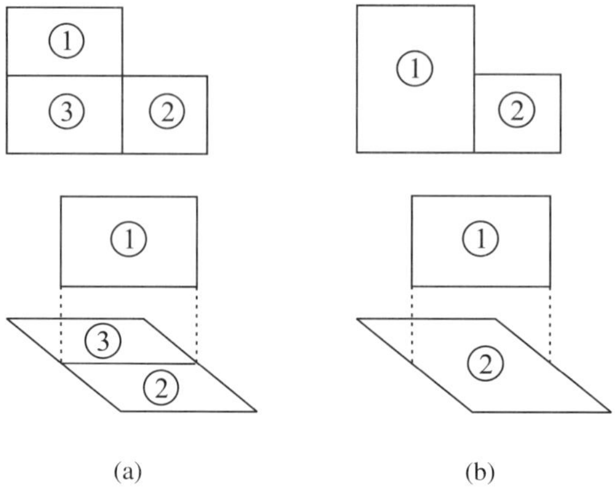

Figure B.2: Surface definition (a) without and (b) with "referencing points".

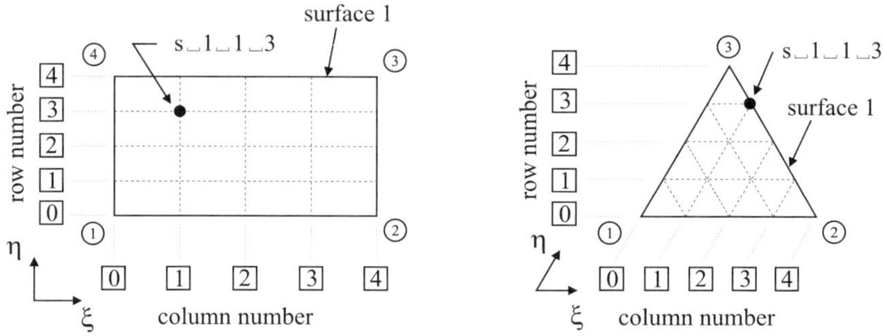

Figure B.3: Referenced node in the first surface entity.

the quadrilateral entities the first integer number specifies the division of the first and the third sides, while the second number specifies the division of the second and the fourth sides. If no more data is specified before the `END` keyword, then the geometric entities are divided uniformly and no one dimensional element will be inserted into the surface or to the edge of the surface.

The lines after the optional `DIVISION_WEIGHT` keyword specify the division weights on the sides. The `weights` can be any real numbers except zero. If the weight is zero, the program will use one as a substitute weight. The specified `weights` are normalised. The `weights` are either specified as a series of double numbers or using the syntax [1] of "`n ( d )`" where the `d` real number is considered `n` times after each other. Any combination of the two syntaxes is also acceptable.

The `LINK_DIRECTION` keyword specifies whether there are one dimensional elements to insert into the surface. The position of the one ('true') value specifies the side with which the one dimensional elements will be parallel (see Figure B.4). The `LINK_DIVISION` and `LINK_TYPE` keywords specify the position and the type of the one dimensional elements. The available types comply with the E_Library specification (as described in Section D.1) which allows five types of one dimensional elements. The positions of the one dimensional elements are given as a series of integer values.

The first integer value in the list specifies the distance from the corresponding side. The distance is expressed in terms of the generated rows of elements. The other values in the list are increments from the previous position. An example is shown in Table B.4. Finally the material can be defined by the `MATERIAL` keyword. The number `n` specifies which user defined material should be assigned to all of the generated surface elements in the surface entity. The number cannot be greater than `NMATERIALS` defined in the header section. The final keyword `DIAGONAL` can be used to control the division of quadrilateral elements into triangles in a quadrilateral geometric entity. The default value is `SHORTEST` which defines that the shortest diagonal of the two diagonals will be used in the geometric entity. The other options allow the use of the longer, first or second diagonal.

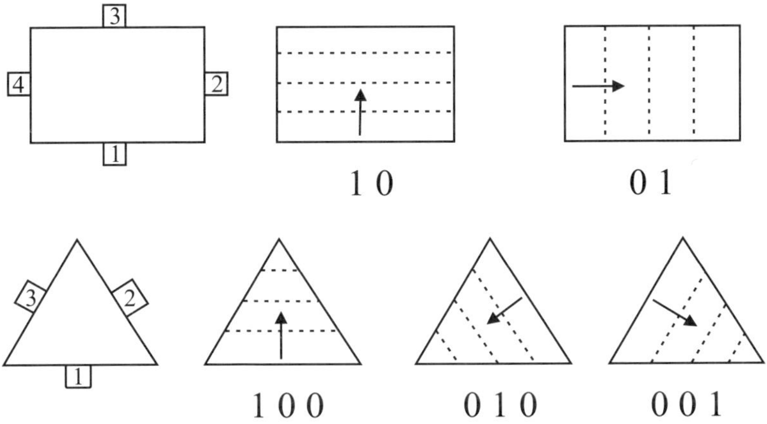

Figure B.4: `LINK_DIRECTION` values and the corresponding direction of the one dimensional elements. Arrows also show the order of the definition of the elements according to `LINK_DIVISION`.

B.3.4 Link entity definition

In this section the one dimensional entities are defined. They can be defined in any order, but their numbering should be consecutive starting from one and the total number of these entities should be equal to `NLINKS` as specified in the header section. These

```
LINK_DIRECTION 1 1
LINK_DIVISION
2 2
0 2 1
LINK_TYPE
3 3
3 5 5
```

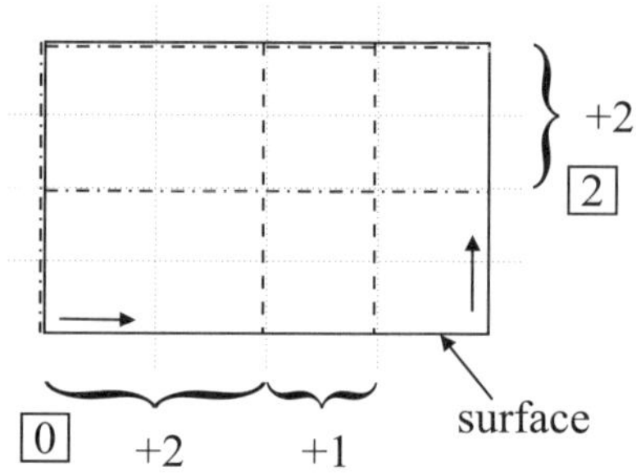

Line	Type
-·-·-·-	3
- - - - -	5

Table B.4: Example for the definition of one dimensional elements in a surface entity.

elements can be defined only as straight lines. The format of the link entity definition is presented in Table B.5. Comments about the `nodes`, element type, division of the element and user definable material can also be applied to this type of element. The difference is that for the one dimensional elements it is not necessary to define `DIVISION`.

```
LINK n
  ELEMENT_TYPE  LINK1 | LINK2 | LINK3 | LINK4 | LINK5
  DEFINITION nodes
  ( N_DIVISIONS n )
  ( DIVISION_WEIGHTS
    weights )
  ( MATERIAL n )
END
```

Table B.5: Link entity definition in the geometric data file.

B.3.5 Boundary condition definition section

The final section contains the boundary condition definitions. Geometrically two types of boundary conditions can be defined: point or line (or area for quadrilateral surfaces). In the final mesh format only point boundary conditions exist, according to the E_Library specification (as described in Section D.1), therefore all boundary conditions are translated to point boundary conditions. The format of the boundary condition definition section is presented in Table B.6. First the point boundary conditions are listed. The point boundary conditions can be defined only for the point entities and they can be listed in any order. The only requirement is that the number of the point boundary conditions should be equal to `NBC_NODES` specified in the header section. The `bc_type` can be “`"FREE"`”, “`"FIXED" d`” or “`"SPRING" d`” where `d` is a real number and expresses the support displacement or the spring constant respectively.

```
BOUNDARY_CONDITIONS
point_entity_index_1   bc_type  bc_type  bc_type
point_entity_index_2   bc_type  bc_type  bc_type
.
.
.
point_entity_index_m   bc_type  bc_type  bc_type

ref_pnt_1  ref_pnt_2   bc_type  bc_type  bc_type
ref_pnt_3  ref_pnt_4   bc_type  bc_type  bc_type
.
.
.
ref_pnt_5  ref_pnt_6   bc_type  bc_type  bc_type
```

Table B.6: Boundary condition definition in the geometric data file.

The boundary condition sides are defined by two “reference points” in an entity and they must lie on a straight line. The “reference point” definition is explained in Section B.3.3. A special case is when the entity is a quadrilateral surface and the two reference points mark an area as is shown in Figure B.5. In this case the same boundary condition will be generated for all points inside the area, including the side and corner points.

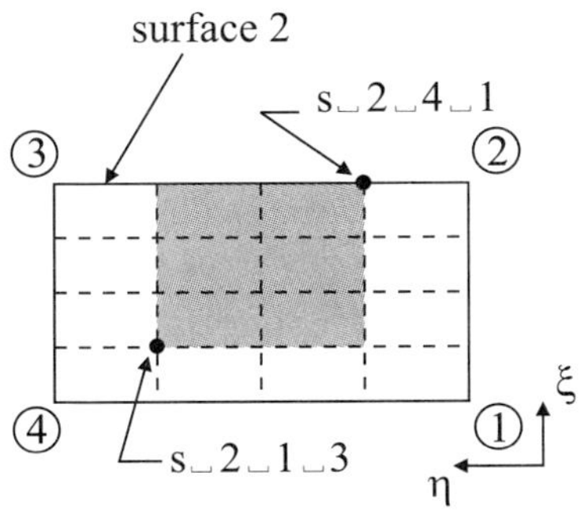

Figure B.5: Boundary condition generation defined by two reference points for an area, the same boundary conditions will be generated for the points inside and on the edge of the shaded area.

First the boundary condition “sides” are generated in the order of their definition. If there are overlapping areas then the boundary conditions of the points in the overlapping area will be redefined and only the later definition will be stored. Finally the boundary condition points are defined and they may redefine the boundary condition for a point again. At every stage of a redefinition of a boundary condition the program issues a warning message. The order of the generation of the boundary conditions and the possibility of the redefinition of previously defined boundary conditions allows that at corners of areas or at the intersection of boundary condition lines, a boundary condition can be defined which may be a combination of the different crossing boundary conditions.

B.4 Examples for parametric definition

An example is shown in Figure B.6. On the left hand side the geometric file definition and on the right hand side the generated mesh is shown. The values of the `PARAMETRIC` definition are determined according to Figure B.7. In Figure B.7(a) the real domain (Ω) is shown. The straight lines determine the quadrilateral surface and the dashed lines are the required shape of the surface. In Figure B.7(b) the unit square domain ($\overline{\Omega}$) is presented. The orientation of the real domain and the unit square domain is the same, in this case it is numbered in an anti-clockwise order. To determine the derivative values for the geometric file which are the components of the tangential vectors in the ξ and η directions in the global coordinate system, the orientation of the ξ,η coordinate system at a particular node should be inspected.

```
NODE_COORDINATES
1 6.0 3.0 0.0
2 6.0 6.0 0.0
3 0.0 0.0 0.0
4 3.0 0.0 0.0

SURFACE 1
  SHAPE    QUAD
  ELEMENT_TYPE   TRIANG3

  DEFINITION PARAMETRIC
  1   0 3 0   -4.71   0      0   0 0 0
  2   0 3 0   -9.42   0      0   0 0 0
  3  -3 0 0    0     -9.42   0   0 0 0
  4  -3 0 0    0     -4.71   0   0 0 0
  N_DIVISIONS 10 10
END
```

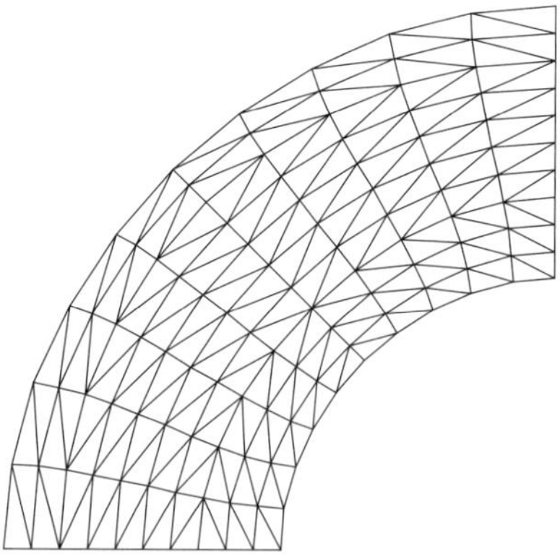

Figure B.6: A complex example to demonstrate the geometric file definition and the generated mesh.

For example, at node 1 in the $\overline{\Omega}$ domain the ξ coordinate direction points from node 1 to node 2, which in the real domain Ω is a vertical vector pointing also from node 1 to

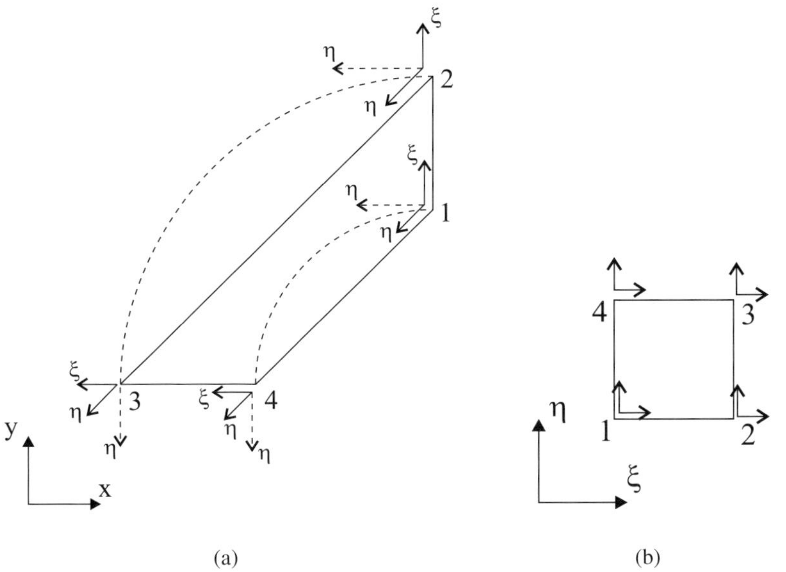

Figure B.7: (a) The real domain Ω and (b) the unit square domain $\overline{\Omega}$ for the complex example.

node 2. The direction of this ξ vector in the global coordinate system is represented in the file as `0 3 0`. The η direction in the $\overline{\Omega}$ domain points from node 1 to node 4. Though in the real domain the vector η also points from node 1 to node 4, however in the real domain the vector should represent the curved edge not a straight edge. Figure B.7(a) shows two η vectors. The η vector with continuous line represents the case when the edge between node 1 and node 4 is straight. On the other hand the η vector with the dashed representation is horizontal to create the curved edge. The line of the η vector is parallel to derivative of the curve or the tangential direction of the curve at node 1. The direction of the η vector is still determined by the direction of the η vector in the $\overline{\Omega}$ domain at node 1. The representation of the η vector at node 1 in the geometric file is `-4.71 0 0`. The length of the vector should be equal to the real side length to ensure equidistant distribution of the points. In this case the quarter length of the perimeter of a circle with a 3 unit radius is $(2r\pi/4 \approx 4.71)$. The last three values for node 1 in the geometric file are zero since there is no "twist" defined. The effect of the "twist" is demonstrated in Table B.13. The method outlined above can be used for all nodes on the surface. However for clarity the η vector at node 4 will be considered. In this case the η vector in the $\overline{\Omega}$ domain points also from node 1 to node 4. To generate a straight edge between node 1 and node 4 the diagonal, continuous η vector should be used as shown in Figure B.7(a). To generate the curved edge, however, the vector which is parallel to the tangential direction of the curve at node 4 should be used. It should be noted, that in a general sense the vector still points from node 1 to node 4 and **not** backwards. The components of this η vector considering the global coordinate system (`0 -4.71 0`) should be put into the geometric file for the fourth node. Another example where the same square surface is defined by the `LINEAR` and the `PARAMETERIC` method is shown in Table B.7. Some other possible variations of the square are shown in Table B.8-B.13. In the tables all examples are in 2D except the last one, which is for a 3D surface.

```
DEFINITION LINEAR
1 2 3 4
```

```
DEFINITION PARAMETRIC
1 1 0 0  0 1 0  0 0 0
2 1 0 0  0 1 0  0 0 0
3 1 0 0  0 1 0  0 0 0
4 1 0 0  0 1 0  0 0 0
```

Table B.7: `LINEAR` and `PARAMETRIC` definition of a square.

Bibliography

[1] PDA Engineering, PATRAN Division, 2975 Redhill Avenue, Costa Mesa, California, USA, *PATRAN Plus User Manual*, October 1990.

[2] Sziveri, J., Seale, C., Editors, *E-lib User's Guide*, SECT Research Group, Heriot-Watt University, Edinburgh, UK, 1997.

[3] Topping, B.H.V., Iványi, P., *Computer Aided Design of Cable-Membrane Structures*, Saxe-Coburg Publications, Stirling, UK, 2004.

```
DEFINITION PARAMETRIC
1 1 1 0  0 1 0  0 0 0
2 1 0 0  0 1 0  0 0 0
3 1 0 0  0 1 0  0 0 0
4 1 0 0  0 1 0  0 0 0
```

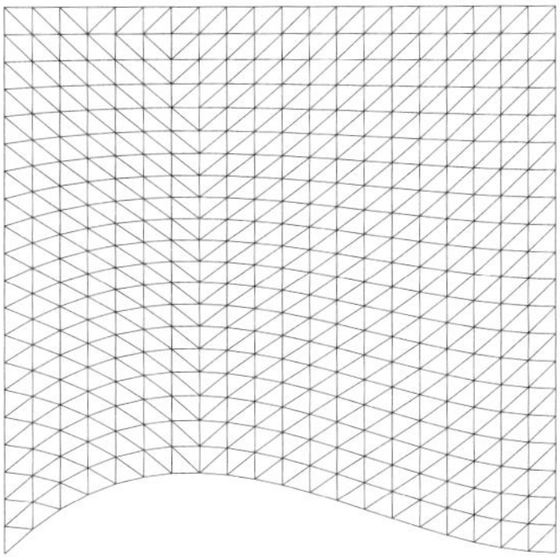

Table B.8: Example for the definition of cubic surfaces.

```
DEFINITION PARAMETRIC
1 1 0 0  1 1 0  0 0 0
2 1 0 0  0 1 0  0 0 0
3 1 0 0  0 1 0  0 0 0
4 1 0 0  0 1 0  0 0 0
```

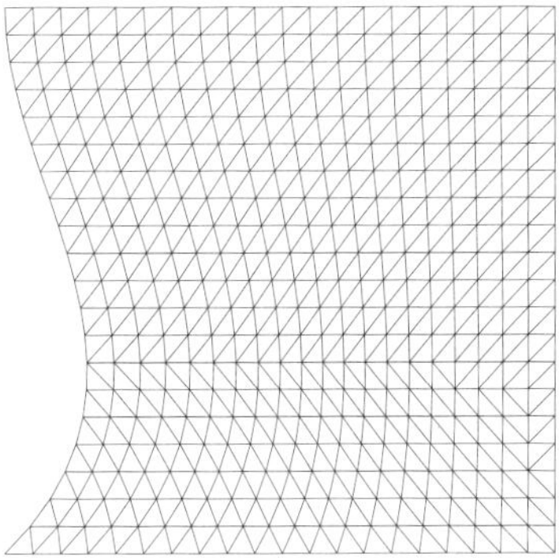

Table B.9: Example for the definition of cubic surfaces.

```
DEFINITION PARAMETRIC
1 1 1 0  1 1 0  0 0 0
2 1 0 0  0 1 0  0 0 0
3 1 0 0  0 1 0  0 0 0
4 1 0 0  0 1 0  0 0 0
```

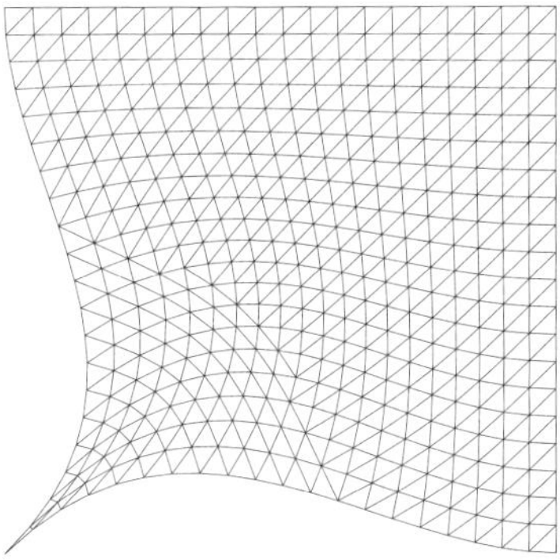

Table B.10: Example for the definition of cubic surfaces.

```
DEFINITION PARAMETRIC
1 1 0 0  0 1 0  10 0 0
2 1 0 0  0 1 0  0 0 0
3 1 0 0  0 1 0  0 0 0
4 1 0 0  0 1 0  0 0 0
```

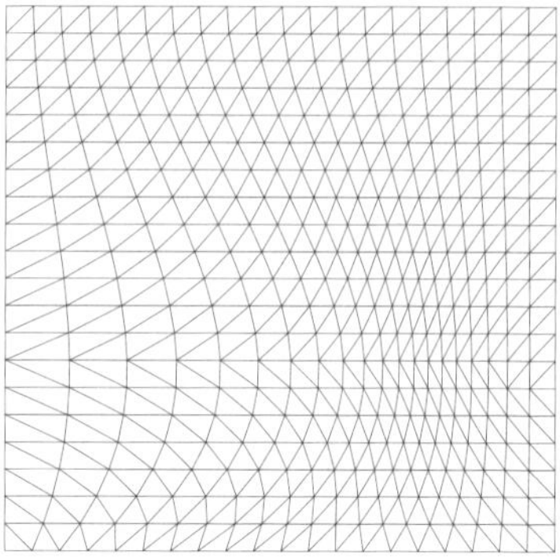

Table B.11: Example for the definition of cubic surfaces.

```
DEFINITION PARAMETRIC
1 1 0 0  0 1 0  10 10 0
2 1 0 0  0 1 0  0 0 0
3 1 0 0  0 1 0  0 0 0
4 1 0 0  0 1 0  0 0 0
```

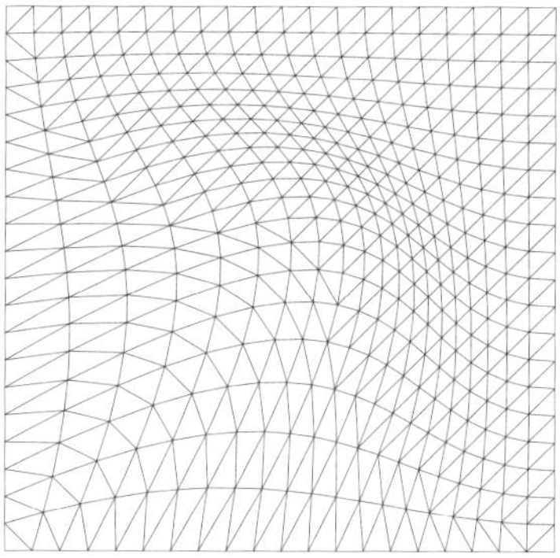

Table B.12: Example for the definition of cubic surfaces.

```
DEFINITION PARAMETRIC
1 1 0 -1 0 1 0  0 0 0
2 1 0 1  0 1 0  0 0 0
3 1 0 1  0 1 0  0 0 0
4 1 0 1  0 1 0  0 0 0
```

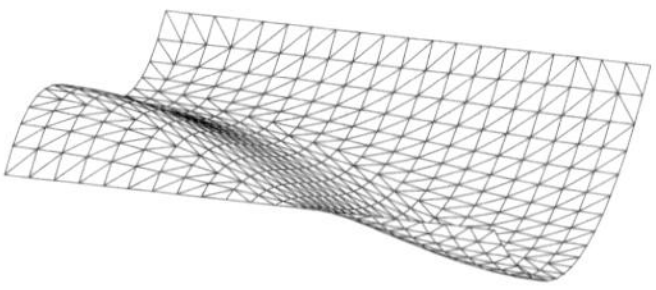

Table B.13: Example for the definition of cubic surfaces.

Appendix C

Mesh generation program input

C.1 The batch file syntax for two dimensional mesh generation

The tasks set for the mesh generator are specified in a batch file. The MGN batch file controls the entire behaviour of the MGN program by listing all the instructions and filenames for the input files. The syntax for the batch file is given in Table C.1 Although this syntax is not incorporated into the E_Library (see Section D.1) it follows the same syntax rules set for all the E_Library files.

C.1.1 Tasks

First the user has to decide what the mesh generator has to do. Four tasks are possible:

- The remeshing of an existing mesh, keeping only the boundaries of the mesh and discarding any existing points inside the domain. If a geometric model is supplied, a new approximation of the boundary will be made. In this case `"REMESH"` should be chosen as a `TASK` setting.

- The refinement of an existing mesh, keeping all existing nodes and refining those existing triangle or quad elements for which the mesh parameter criteria is not satisfied. In this case `"REFINE"` should be chosen as a `TASK` setting.

- The refinement of an existing mesh on an element by element basis. In this case `"EBE_REFINE"` should be chosen as a `TASK` setting.

- The post-processing or smoothing of an existing mesh is possible when `TASK` is set to `"POSTPROC"`.

TASK	"REMESH" | "REFINE" | "EBE_REFINE" | "POSTPROC"
METHOD	"ADV_FRONT" | "DELAUNAY" | "PAVING" | "REG_GRID" | "NONE"
ALFA	d
BETA	d
DLN_OPTION	"CENTROID" | "CENTR_RAD" | "PROJECTION"
CONTROL_TYPE	"BACKGROUND" | "MPR" | "SOURCES" | "UNIFORM"
MESH_OPTION	"FISSION_TO_QUAD" | "RANK_TO_QUAD" | "HALF_TO_QUAD" | "SPLIT_TO_TRIANG" | "NONE"
GEOM_MODEL_FILE	"*"
COARSE_MESH_FILE	"*"
RESULT_MESH_FILE	"*"
BKGRND_MESH_FILE	"*"
BKGRND_MESH_PARAMS_FILE	"*"
COARSE_MESH_PARAMS_FILE	"*"
LOCAL_MESH_CTRL_FILE	"*"
UNIFORM_MESH_PARAM	d
SMOOTHING_TYPE	"LAPLACIAN" | "LAPLACIAN_NDE" | "NODAL_AVG" "PAVING" | "NONE"
SMOOTHING_ITER	n
OUTPUT_FORMAT	"E_LIB"

Table C.1: Syntax of the **mgn** batch file.

C.1.2 Methods

The mesh generator supports a number of methods for generating unstructured meshes:

- The advancing front technique which can be slow, but creates very regular high quality meshes. In this case "ADV_FRONT" should be selected as METHOD setting.
- The Delaunay triangulation algorithm (by point insertion) which is on average ten times faster than the advancing front method, but for which regularity and quality of the mesh can be less than by the advancing front method. In this case "DELAUNAY" should be chosen as a METHOD setting.
- The paving method is a good quality, direct quadrilateral mesh generator method. In this case "PAVING" should be chosen as a METHOD setting.
- For an experimental regular triangle grid generation method the METHOD should be set to "REG_GRID". This method is not described in the book, however the reader is directed to reference [2] for details of this method.
- When the task is post-processing ("POSTPROC") the METHOD should be set to NONE.

When the Delaunay triangulation is chosen, the user has to specify two extra parameters ALFA and BETA.

- Alfa controls the conformity of the mesh density with the parameters set by the control type. When DLN_OPTION is "CENTROID" the requirement for alfa is $1.0 < \alpha$. When DLN_OPTION is "CENTR_RAD" the requirement for alfa is $0.5 \leq \alpha \leq 2.0$. Values close to 1.0 have been found to be a good starting point.
- Beta ($0.1 \leq \beta \leq 15.0$) controls the regularity of the mesh. Lower values of beta generate more chaotic meshes whereas higher values would give a more regular mesh. A good default value is 2.0.

C.1.3 Delaunay options

When the Delaunay triangulation is chosen the user must define which Delaunay refinement strategy is to be used. This is controlled by the DLN_OPTION keyword. The following options are implemented:

- "CENTROID" controls the mesh size on the basis of side lengths of triangles. Centroidal nodes are inserted into a triangle when at least one of its side lengths is longer than allowed.
- "CENTR_RAD" controls the mesh size on the basis of the lengths of the radial connection between the centroid and the corner node of triangles. The centroid node is inserted into a triangle when the radial connection is longer than allowed.
- "PROJECTION" behaves like "CENTROID" but it uses projective point placement instead of centroid nodes.

C.1.4 Control types

An efficient control of the mesh density at any place inside the domain and on the borders is essential for any good mesh generator. The implemented mesh generator offers four control types which have also been discussed in Chapter 3:

- The use of a background mesh that 'sits' behind the domain is a common technique. For every new point to be generated in the mesh the mesh parameter is derived from the value this point would have if it belonged to the background mesh. Therefore care has to be taken that the background mesh covers the complete area of the domain, including potential extensions of the domain by a new approximation of the boundary curves. The background mesh also has to be convex. In this case "BACKGROUND" should be chosen as a CONTROL_TYPE setting and the BKGRND_MESH_FILE and the BKGRND_MESH_PARAMS_FILE keywords must also be defined.

- Mesh parameter controlled meshing is equivalent to background controlled meshing where the background mesh is the coarse mesh itself. However the mesh parameters specified in the COARSE_MESH_PARAMS_FILE file have to be nodal mesh parameters. (If element mesh parameters are specified they will be converted into nodal parameters by averaging.) In this case "MPR" should be included as a CONTROL_TYPE and the COARSE_MESH_PARAMS_FILE keyword must also be defined.

- The use of local control sources is the third type. It is an extension to the background controlled meshing whereby the user can define some local control sources in a separate file that will influence the mesh density locally. Section C.3 explains the syntax and the effect of these sources. In this case "SOURCES" should be selected as a CONTROL_TYPE setting. In addition the keywords for LOCAL_MESH_CTRL_FILE must also be defined.

- The uniform mesh parameter is the simplest control type. In this case, of course, the size of the generated elements is uniform and the UNIFORM_MESH_PARAM keyword should also be defined.

C.1.5 Mesh options

A triangular mesh generated using the Delaunay or advancing front method can automatically be converted into a quadrilateral mesh, while a quadrilateral mesh generated by the paving method can be converted into a triangular mesh. The following conversion options are supported:

- "FISSION_TO_QUAD": is a method of converting triangular meshes into quad meshes by melting two triangles to form one quad. However the method also considers and avoids individual remaining triangles as discussed in Section 8.3.

- "RANK_TO_QUAD": is a method to convert triangle meshes into quad meshes. The method was developed by Rank *et al* [3] and discussed in Section 8.2.

- "HALF_TO_QUAD": splits each triangle into three quads by using points at the middle of all edges of the triangle and a point in the centre of the triangle.

- "SPLIT_TO_TRIANG": converts quadrilateral meshes into triangular meshes by splitting every quadrilateral into two triangles along the shortest diagonal.

- "NONE" should be used if no element conversion is necessary.

C.1.6 Filenames

Filename references have to be absolute or relative to the location of the executable. Filenames are always given without extension as it is assumed that the extension matches the filetype syntax. Filenames are enclosed in double quotes. The following filenames may be defined:

- `GEOM_MODEL_FILE`: the location and name of the geometric model file without the `*.gmf` extension. This file is required for new mesh generation from geometric model only. It may be, but does not have to be supplied for the other tasks.
- `COARSE_MESH_FILE`: the location and name of the coarse mesh definition file without the `*.mdf` extension.
- `RESULT_MESH_FILE`: the location and name of the resulting mesh to which an extension will be added according to the output format. If the file exists already, it will be overwritten. This filename always has to be supplied.
- `BKGRND_MESH_FILE`: the location and name of the background mesh definition file without the `*.mdf` extension. This file is required when background or source controlled mesh generation is selected.
- `BKGRND_MESH_PARAMS_FILE`: the location and name of the background mesh parameters file without the `*.mpr` extension. This file is required when background or source controlled mesh generation is selected.
- `COARSE_MESH_PARAMS_FILE`: the location and name of the mesh parameters file without the `*.mpr` extension. This file is required when mesh paramater controlled mesh generation is chosen.
- `LOCAL_MESH_CTRL_FILE`: the location and name of the local control sources file without the `*.ctl` extension. This file is required when local control sources are specified.

C.1.7 Smoothing

The mesh that results from any of the operations mentioned above can be optimized or smoothed. The only smoothing type that is currently supported is optimized Laplacian smoothing. In this case the `"LAPLACIAN"` or the `"LAPLACIAN_NDE"` keyword must be chosen. The `"LAPLACIAN"` keyword means a Laplacian smoothing with diagonal exchange, while the `"LAPLACIAN_NDE"` keyword denotes a Laplacian smoothing without diagonal exchange. `"NODAL_AVG"` will activate only nodal averaging. If `"NONE"` is chosen as a `SMOOTHING_TYPE` setting no smoothing is carried out at all. If smoothing is activated, the number of smoothing iterations to be carried out must be specified . This number is set by the `SMOOTHING_ITER` keyword and has to be greater than one. Two iterations are in most cases sufficient.

The `"PAVING"` smoothing type must however be specified if `"PAVING"` is chosen as the mesh generation method, as a result of the internal procedures of the paving method.

NMESHES	n
INPUT_MESH_FILES	"*" "*" ...
OUTPUT_MESH_FILE	"*"
REFINE_METHOD	"EDGE_LEN"
ALPHA	d
BETA	d
LOG_LEVEL	"FULL_LOG" | "NORMAL_LOG" | "MINIMAL_LOG" | "NO_LOG"
OPTIMIZATION	"INSERTING" | "SWAPPING" | "LAPLACE" | "COLLAPSING"

Table C.2: Syntax of the **ctr** batch file.

C.1.8 Output format

The output format of the resulting mesh can be specified by `OUTPUT_FORMAT` keyword. Currently `"E_LIB"` and `"SMSH"` are supported. The E_Library output will add the `*.mdf` extension to the resulting mesh filename.

C.2 The batch file syntax for the three dimensional mesh generation

The CTR batch file controls the behaviour of the three dimensional mesh generator program. The syntax for the batch file is given in Table C.2

The `NMESHES` keyword defines the number of input meshes provided after the `INPUT_MESH_FILES` keyword. The first mesh supplied for this keyword must enclose all other meshes. The output result of the mesh generation will be written into the file given by the `OUTPUT_MESH_FILE` keyword. The type of mesh refinement method is specified using the `REFINE_METHOD` keyword. Only one type of mesh refinement method is available: `''EDGE_LEN''`. The specification of the `''EDGE_LEN''` method means that the mesh parameters are compared with the edges of the tetrahedra elements. `ALPHA` and `BETA` are two parameters for the Delaunay Kernel as described in Section 7.9.8. The `LOG_LEVEL` keyword specifies how much information about the mesh generation process should be stored in a log file. Specifying `''FULL_LOG''` requires all available information to be stored in the log file. The `OPTIMIZATION` keyword determines which methods are to be used to improve the quality of the final mesh. `INSERTING` denotes the method where new points are inserted into the triangulation in the circum-sphere of "bad" elements, therefore the "bad" quality element will be removed. `SWAPPING`, `LAPLACE` and `COLLAPSING` operates on a mesh which is not necessarily a Delaunay mesh. The `SWAPPING` keyword denotes the method of edge and face swapping (see Section 9.7.3.1). The `LAPLACE` keyword selects a Laplacian smoothing method (see Section 9.8.1), while the `COLLAPSING` keyword denotes edge collapsing.

```
NCONTROL_SETS        2

NCONTROL_POINTS      2
  COORDINATES        8.5 2.0 0.0
  INNER_RADIUS       0.2
  OUTER_RADIUS       1.0
  MESH_PARAMETER     0.3

  COORDINATES        1.5 2.0 0.0
  INNER_RADIUS       0.2
  OUTER_RADIUS       1.5
  MESH_PARAMETER     0.2

NCONTROL_POINTS      1
  COORDINATES        2.0 5.5 0.0
  INNER_RADIUS       0.2
  OUTER_RADIUS       1.0
  MESH_PARAMETER     0.2
```

Table C.3: An example local control file (**ctl**) with a line-wise and a point-wise source.

C.3 Local control source file syntax

The local mesh density control data file (`.ctl`) defines a set of local control points for point-wise and line-wise mesh size control as discussed in Section 3.3.3. For point-wise local control the point coordinates, the radius of the inner circle, the radius of the outer circle and the nodal mesh size parameter are specified. For line-wise local control a chain of control points must be specified. The syntax is explained by an example file, shown in Table C.3, where first a line-wise size control and then a point-wise control is defined.

C.4 Geometric definition file syntax

There are two most common nonlinear mathematical forms in geometric modelling for curve and surface representation, one is implicit and another is the parametric polynomial forms. The implicit form has the advantage that circles, conics, and primitive quadric surfaces, such as cylinders, spheres and cones can be concisely and precisely represented. A disadvantage of the implicit form is that free-form curves and surfaces, which are also important in geometric modelling, cannot be represented. With parametric polynomials, such as polynomial B-splines, one can represent and manipulate free-form curves and surfaces; but unfortunately, circles, conics and the quadric primitives cannot be represented precisely. Non-uniform Rational B-spline (NURBS) offers the advantages of both forms.

The use of rational polynomial functions for representing curves and surfaces in CAD/CAM applications is becoming increasingly widespread. NURBS have been an Initial Graphics Exchange Specification(IGES) standard [1] since 1983. A large number of inhouse and commercial modelling systems have been developed in CAD/CAM using NURBS. In order to keep compatibility with all those systems the NURBS representation has been adopted as a standard here.

The element geometric definition file (`.gmf`) defines the boundary of the problem domain using Non Uniform Rational B-Spline (NURBS). The file contains the following keywords:

- NENDPOINTS: Number of control points used for the geometric modelling. The boundary nodes on the current mesh can be employed as control points. They will be referred to by their node numbers in the curve specification.
- NNURBS_CURVES: Number of NURBS curves.
- ENDPOINT_COORDINATES: A list of coordinates of end points.
- NURBS_CURVE: Defines each NURBS curve.
- DEGREE: The number of degrees of freedom for the NURBS curve.
- CONTROL_POINTS: The four control points for cubic NURBS curve.
- WEIGHTS: The value of weights for four control points.

The geometric definition file should only be used together with the remeshing data extensions for the mesh definition file. These extensions are not covered in the main definition of the **mdf** syntax and can be best explained by the example shown in Figure C.1.

Figure C.1(a) shows one coarse triangle defined by nodes 1, 2 and 3. On the side between nodes 1 and 2 a NURBS curve `C1` is defined. The curve is cubic (degree 3) and connects node 1 where it has parameter 0.0 to node 2 where it has parameter 1.0. Two control points determine the shape of the curve. These control points have coordinates (0, -10, 0) and (10, -10, 0). The weights for these control points are set to 0.5. Figure C.1(b) shows the remeshed triangle when the three nodal mesh parameters were defined with value $\delta = 0.8$. Table C.4 and C.5 shows the **mdf** and the **gmf** files for the example in Figure C.1

Bibliography

[1] Association, N.C.G., "Initial Graphics Exchange Specification (IGES) Version 5.3", Technical report, IGES/PDES Organization, 2722 Merrilee Drive, Suite 200, Fairfax, VA., USA, 1995.

```
TITLE "coarse mesh"
NMESHPOINTS      3
NNODES      3
NELEMENTS_TRIANG1    1
NNURBS_CURVES        1

MESHPOINT_COORDINATES
1    0.0      0.0     0.0
2    10.0     0.0     0.0
3    5.0      8.67    0.0

NODES_TRIANG1
1    1     2     3

REMESH_DATA
1    "EXTERIOR"
     NNURBS_CURVES    1
     1    NURBS_CURVE "C1"
          PARAMETER   0.0
2    "EXTERIOR"
     NNURBS_CURVES    1
     1    NURBS_CURVE "C1"
          PARAMETER   1.0
3    "EXTERIOR"
```

Table C.4: The **mdf** file for the example shown in Figure C.1.

```
NENDPOINTS  2
ENDPOINT_COORDINATES
  1    0.0  0.0   0.0
  2    10.0 0.0   0.0

NURBS_CURVE   "C1"
DEGREE   3
CONTROL_POINTS
  1    1
  2    0.0  -10.0  0.0
  3    10.0 -10.0  0.0
  4    2
WEIGHTS
  1    1.0
  2    0.5
  3    0.5
  4    1.0
END
```

Table C.5: The **gmf** file for the example shown in Figure C.1.

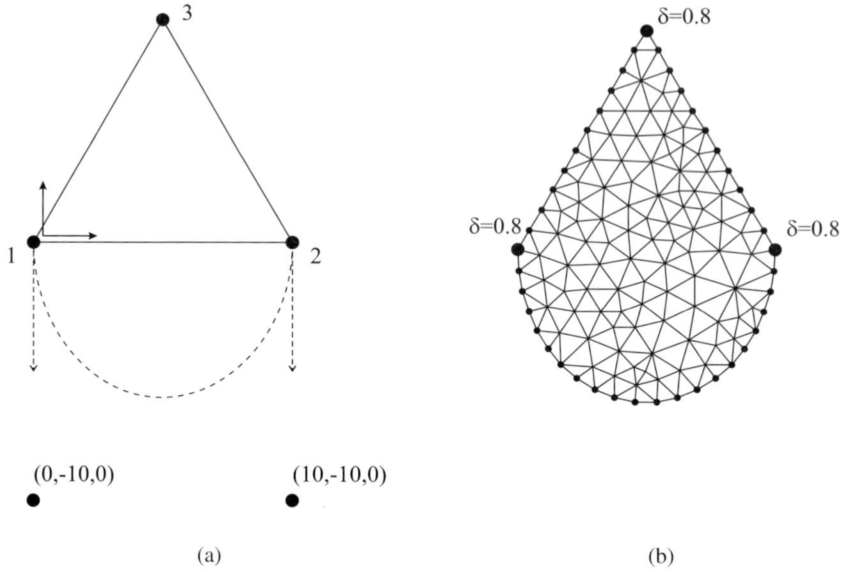

Figure C.1: An example for the geometric definition.

[2] Muylle, J., Iványi, P., Topping, B.H.V., "A new point creation scheme for uniform Delaunay triangulations", *Engineering Computations, International Journal for Computer Aided Engineering and Software*, 19(6): 707–735, 2002.

[3] Rank, E., Schweingeruber, M., Sommer, M., "Adaptive mesh generation and transformation of triangular to quadrilateral meshes", *Commmunications in Numerical Methods in Engineering*, 9: 121–129, 1993.

Appendix D

Computational Tools

D.1 The E_Library

This section is based on an early manual, written by János Sziveri and Colin Seale, for the E_Library [1]. They were both members of the Structural Engineering Computational Technology Research Group, at Heriot-Watt University, Edinburgh, and the first developers responsible for the development of this library.

D.1.1 Introduction

The main motivation behind the development of this library was to provide a common data format and functionality which can be used and understood by all components of an application involved in pre-processing, finite element methods and post-processing. The requirements of the E_Library format were that it should describe all important aspects of the finite element meshes in an easily understood format. Furthermore it should be simple to use and implement. The full description of the E_Library format can be found in reference [1]. The E_Library data format is an ASCII text file which provides 'manual' access to the data "if everything else fails". The files can contain one or several lines of comments and an arbitrary number of spaces. The data itself is organised by keywords. Each section of data is preceded by a keyword. This format increases the readability of the data and it also provides a means of data validation. Some keywords are mandatory while other keywords are either optional or data dependent. There are no physical dimensions or units in the data files. This is left to the programmer or the user to implement or use the appropriate values and conversions of units. The mesh description data and other related data is stored in different files which are identified by the extension names. The `mdf` file contains the geometric and topological description of the mesh. It also contains reference data to other files, for example material references. The `mat` file provides the material definitions, the `dom` file contains the subdomain decomposition information, the `mpr` file has the mesh parameters, the `gmf` file extends the description of the geometry of the mesh with curve definitions while the `ste` and `fee` files contain the resulting stresses and finite element errors. The library also provides a programming interface by defining the data structures and functional interface to these data structures. The main collective structure is the `MESH` structure which gathers all data related to a particular problem. This structure has several sub-structures embedded into it which are separated into different files. The functional interface enables the reading-in, writing-out of these structures. It also provides an integrity check on the data, especially during the reading-in phase. There are several other supporting functions, for example memory management, enquiring functions and functions to distribute the data to different processes in a parallel execution environment.

The library is implemented in ANSI C and it has been successfully compiled under several operating systems such as: MS-DOS, Windows 95, Windows NT, Solaris, IRIX and Linux.

D.1.2 E_Library file format

The E_Library format supports meshes of a single element type and mixed meshes. The E_Library mesh file format takes the form of a series of keywords followed by one or more data items. For example:

```
NELEMENTS_TRIANG1    120
```

denotes that the mesh contains 120 elements of type **TRIANG1**. Most keywords are optional, but those which are declared must be in a strict, predefined order. This order automatically defines the dependencies between the data items.

The format also allows the use of file comments. If the first non-white character on a line is the # character, then all remaining text on that line is ignored. Two or more lines can be commented out by enclosing the lines inside the symbols /* and */. The opening /* must be the first two characters on the first comment line and */ the first two characters on the last comment line. All text on the line following /* and before */ is ignored. The /*...*/ comments must not be nested, although they can contain # comments. Blank lines are also permitted.

There are seven types of data files – the mesh definition (**.mdf**), geometric definition (**.gmf**), materials (**.mat**), domain decomposition (**.dom**), element mesh parameters (**.mpr**), stresses (**.ste**), finite element errors files (**.fee**). The mesh definition file contains the main description of the mesh. The other files describe additional properties of the mesh or they represent a state of the mesh.

The following sections describe the keywords and keyword data for the above data files. Each section contains a table which lists the keywords in the order that they must be declared and defines the associated keyword data. The data type – whether it is an integer (i), real (r) or a character string (s) – is also given, where, for example, 'i 3*r' denotes that the data consists of an integer followed by three reals. Units, where relevant, are enclosed in square brackets ([...]).

Keyword data consisting of a single data item must follow the keyword on the same line. For vectors and matrices, each vector component or matrix row must start on a new line and each line must begin with an integer index. Unless otherwise stated, this index is either the vector component number or matrix row index. All keyword data strings must be enclosed in double quotes.

D.1.3 The mesh definition file

The mesh definition file (**.mdf**) contains the main description of the mesh. Twelve different elements types are supported – five one dimensional (**LINK1**, **LINK2**, **LINK3**, **LINK4**, **LINK5**), five triangular (**TRIANG1**, **TRIANG2**, **TRIANG3**, **TRIANG4**, **TRIANG5**), two quadrilateral (**QUAD1** and **QUAD3**), one tetrahedral (**TETRAH1**) and one three dimensional block element (**BLOCK1**). These element types are defined in Figure D.1 which shows the order in which the element nodes must be labelled. Table D.1 gives a short description of the elements. With the exception of the **TRIANG2** and **QUAD3** elements, individual elements have uniform material properties. **TRIANG2** and **QUAD3** elements are composed of layered composite materials referred to as composite materials of type 1.

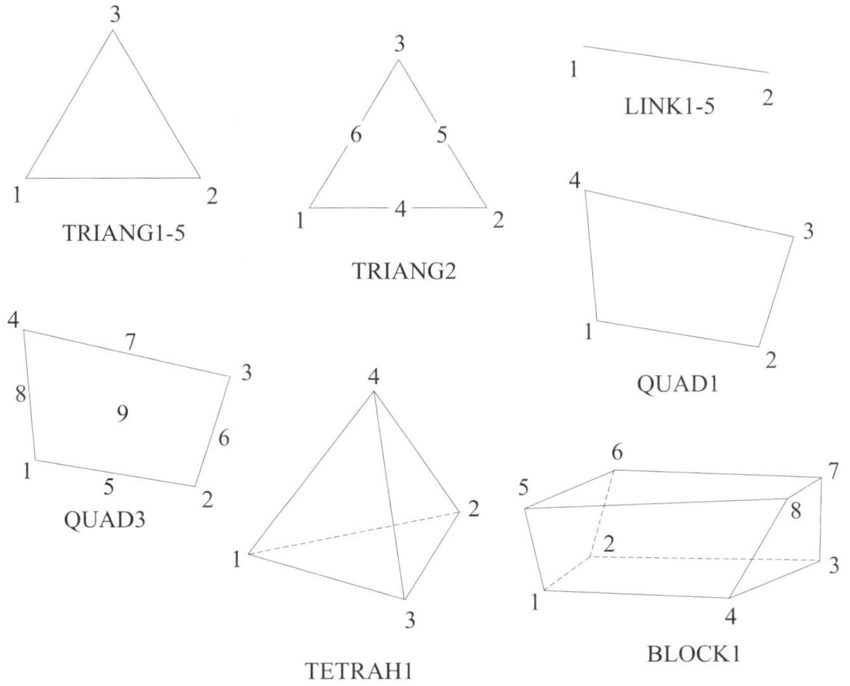

Figure D.1: The element node orders.

Element Type	Description	Number of Vertices	Number of Nodes
LINK1	Truss	2	2
LINK2	Cable	2	2
LINK3	Fixed tension 1-D link	2	2
LINK4	Fixed force density 1-D link	2	2
LINK5	Geodesic string	2	2
TRIANG1	Constant strain triangular	3	3
TRIANG2	A combined constant strain plane stress and constant moment plate bending simple facet triangular	3	6
TRIANG3	Solid triangular	3	3
TRIANG4	Membrane triangular	3	3
TRIANG5	Constant stress triangular	3	3
QUAD1	Plane stress quadrilateral	4	4
QUAD3	Mindlin plate quadrilateral	4	9
TETRAH1	Tetrahedral	4	4
BLOCK1	Block	8	8

Table D.1: The element types.

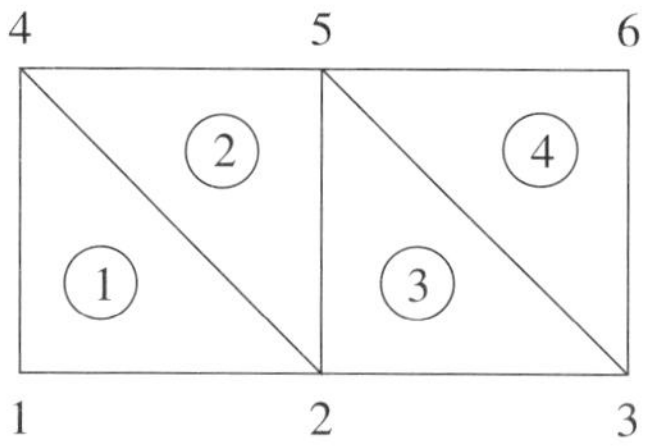


```
# An example mesh definition file

TITLE "An example mesh"
NMESHPOINTS          6
NNODES               6
NELEMENTS_TRIANG1    4

MESHPOINT_COORDINATES
  1     0.000      0.000      0.000
  2     3.000      0.000      0.000
  3     6.000      0.000      0.000
  4     0.000      3.000      0.000
  5     3.000      3.000      0.000
  6     6.000      3.000      0.000

NODES_TRIANG1
  1       1          2          4
  2       2          5          4
  3       2          3          5
  4       3          6          5
```

Table D.2: An example with TRIANG1 elements.

An example mesh definition file is shown in Table D.2. The first keyword **TITLE**, is compulsory and specifies the title of the mesh. The title can be used to annotate output such as the display in a plot program and, as with all string arguments, must be enclosed in double quotes.

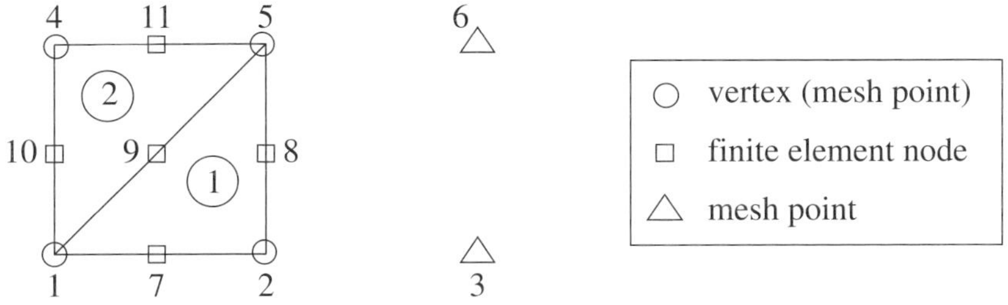

Figure D.2: An example with TRIANG2 elements.

```
# Example with TRIANG 2 elements

TITLE "An example mesh"
NMESHPOINTS            6
NNODES                 10
NELEMENTS_TRIANG2      2

MESHPOINT_COORDINATES
  1     0.000     0.000     0.000
  2     3.000     0.000     0.000
  3     6.000     0.000     0.000
  4     0.000     3.000     0.000
  5     3.000     3.000     0.000
  6     6.000     3.000     0.000

NODES_TRIANG2
  1     1         2         5         7         8         9
  2     1         5         4         9        11        10
```

Table D.3: Definition of the example shown in Figure D.2.

A mesh is defined in terms of a set of *mesh-points* which are joined by straight lines to form the edges or surfaces of the elements. A distinction is made between the terms *mesh-point*, *vertex* and *finite element node*. A mesh-point is a point in space which may or may not form an element vertex, whereas a finite element node (or simply *node*) is a point on an element used for function interpolation during the finite element analysis. Unlike mesh-points, nodes need not coincide with the element vertices. (See Figure D.2.) **NMESHPOINTS** is the keyword for the number of mesh-points defined in the mesh definition file and **NNODES** defines the total number of finite element

```
# Example with extra keywords

TITLE "An example mesh"
NMESHPOINTS                  3
NNODES                       3
NELEMENTS_TRIANG1            1
NBOUNDARY_CONDITION_NODES    2

MESHPOINT_COORDINATES
  1     0.000     0.000     0.000
  2     6.000     0.000     0.000
  3     6.000     6.000     0.000

NODES_TRIANG1
  1      1         2         3

BOUNDARY_CONDITIONS
  1    "FIXED" 0.000    "FIXED" 0.000    "FREE"
  2    "FREE"           "FIXED" 0.000    "FREE"
```

Table D.4: Definition of an example with boundary conditions.

nodes in the mesh. The coordinates of the mesh-points follow the keyword **MESHPOINT_COORDS**. All three keywords **NMESHPOINTS**, **NNODES** and **MESHPOINT_COORDS** are compulsory. The keyword **NELEMENTS_TRIANG1** act both to declare the type of elements in the mesh and to give the number of the element. If this number is zero then the keyword should be omitted. At least one of the keywords which defines the element type must be declared and for each such keyword present there must also be the corresponding keyword such as **NODES_TRIANG1** etc. The data for these keywords defines the node indices of each element in turn in the order given in Figure D.1. Element vertex nodes are always labelled before non-vertex nodes and for two dimensional elements, labelling occurs in an anti-clockwise direction.

Another example is presented to demonstrate the difference between *mesh-point*, *vertex* and *finite element node*. The geometric arrangement of the elements can be seen in Figure D.2 while the generated mesh file is shown in Table D.3. *All other keywords are optional* except in cases where the declaration of one keyword implies that one or more further keywords will be declared later. For example, if the number of boundary condition nodes is defined (**NBOUNDARY_CONDITION_NODES**) then the nodal boundary conditions (**BOUNDARY_CONDITIONS**) must also be defined. See Table D.4 for an example.

An exhaustive list of possible keywords in an **.mdf** file is presented in Tables D.5(a), (b) and (c). For the use of keywords not discussed here see Reference [1].

Keyword	Keyword Data	Data Type
TITLE	Mesh title	s
NMESHPOINTS	Number of mesh-points	i
NNODES	Number of nodes	i
NELEMENTS_LINK1	Number of **LINK1** elements	i
NELEMENTS_LINK2	Number of **LINK2** elements	i
NELEMENTS_LINK3	Number of **LINK3** elements	i
NELEMENTS_LINK4	Number of **LINK4** elements	i
NELEMENTS_LINK5	Number of **LINK5** elements	i
NELEMENTS_TRIANG1	Number of **TRIANG1** elements	i
NELEMENTS_TRIANG2	Number of **TRIANG2** elements	i
NELEMENTS_TRIANG3	Number of **TRIANG3** elements	i
NELEMENTS_TRIANG4	Number of **TRIANG4** elements	i
NELEMENTS_TRIANG5	Number of **TRIANG5** elements	i
NELEMENTS_QUAD1	Number of **QUAD1** elements	i
NELEMENTS_QUAD3	Number of **QUAD3** elements	i
NELEMENTS_TETRAH1	Number of **TETRAH1** elements	i
NELEMENTS_BLOCK1	Number of **BLOCK1** elements	i
NBOUNDARY_CONDITION_NODES	Number of boundary condition nodes	i
NLOADED_NODES	Number of loaded nodes	i
NMATERIALS	Total number of materials including those declared in composite materials	i
NCOMP_MATERIALS_TYPE1	Number of type 1 composite materials	i
NNURBS_CURVES	Number of NURBS curves	i
TIMESTEP	The value of the timestep	r
NTIMESTEPS	Number of time-steps This keyword and the pair of keywords **NINTERNAL_TIMESTEPS** and **NEXTERNAL_TIMESTEPS** are mutually exclusive	i
NINTERNAL_TIMESTEPS **NEXTERNAL_TIMESTEPS**	Number internal and external time-steps This data enables time-stepping to be broken down into a series of *NExternalTimesteps* sets of *NinternalTimesteps* time-steps (See **NTIMESTEPS**)	i
DAMPING_FACTOR	Viscous damping factor	r
BETA	Newmark's β integration constant	r
GAMMA	Newmark's γ integration constant	r

Table D.5: (a) The mesh definition file keywords and keyword data (cont.)

Keyword	Keyword Data	Data Type
NSTRESS_POINTS_LINK1	Number of stress points in a LINK1 element	i
NSTRESS_POINTS_LINK2	Number of stress points in a LINK2 element	i
NSTRESS_POINTS_LINK3	Number of stress points in a LINK3 element	i
NSTRESS_POINTS_LINK4	Number of stress points in a LINK4 element	i
NSTRESS_POINTS_LINK5	Number of stress points in a LINK5 element	i
NSTRESS_POINTS_TRIANG1	Number of stress points in a TRIANG1 element	i
NSTRESS_POINTS_TRIANG2	Number of stress points in a TRIANG2 element	i
NSTRESS_POINTS_TRIANG3	Number of stress points in a TRIANG3 element	i
NSTRESS_POINTS_TRIANG4	Number of stress points in a TRIANG4 element	i
NSTRESS_POINTS_TRIANG5	Number of stress points in a TRIANG5 element	i
NSTRESS_POINTS_QUAD1	Number of stress points in a QUAD1 element	i
NSTRESS_POINTS_QUAD3	Number of stress points in a QUAD3 element	i
NSTRESS_POINTS_TETRAH1	Number of stress points in a TETRAH1 element	i
NSTRESS_POINTS_BLOCK1	Number of stress points in a BLOCK1 element	i
MESHPOINT_COORDS	x-,y- and z-coordinates of the mesh-point	i 3*r
NODES_LINK1	Node indices of each **LINK1** element	i 2*i
NODES_LINK2	Node indices of each **LINK2** element	i 2*i
NODES_LINK3	Node indices of each **LINK3** element	i 2*i
NODES_LINK4	Node indices of each **LINK4** element	i 2*i
NODES_LINK5	Node indices of each **LINK5** element	i 2*i
NODES_TRIANG1	Node indices of each **TRIANG1** element	i 3*i
NODES_TRIANG2	Node indices of each **TRIANG2** element	i 6*i
NODES_TRIANG3	Node indices of each **TRIANG3** element	i 3*i
NODES_TRIANG4	Node indices of each **TRIANG4** element	i 3*i
NODES_TRIANG5	Node indices of each **TRIANG5** element	i 3*i
NODES_QUAD1	Node indices of each **QUAD1** element	i 4*i
NODES_QUAD3	Node indices of each **QUAD3** element	i 9*i
NODES_TETRAH1	Node indices of each **TETRAH1** element	i 4*i
NODES_BLOCK1	Node indices of each **BLOCK1** element	i 8*i
BOUNDARY_CONDITIONS	Nodal boundary conditions For each boundary condition node, the node index followed by, for each degree of freedom, the string **"FREE"** if no boundary conditions are imposed, **"FIXED"** followed by a real displacement [m], or **"SPRING"** followed by a non-negative spring-constant [Nm^{-1}]. The displacement argument specifies an initial displacement of a node. The spring constant enables elastic forces to be modelled.	←

Table D.5: (b) The mesh definition file keywords and keyword data (cont.)

Keyword	Keyword Data	Data Type
LOADS	Nodal loads For each loaded node, the node index followed by the applied load [N] for each degree of freedom	i r
MATERIALS_LINK1	Material name of each **LINK1** element	i s
MATERIALS_LINK2	Material name of each **LINK2** element	i s
MATERIALS_LINK3	Material name of each **LINK3** element	i s
MATERIALS_LINK4	Material name of each **LINK4** element	i s
MATERIALS_LINK5	Material name of each **LINK5** element	i s
MATERIALS_TRIANG1	Material name of each **TRIANG1** element	i s
COMP_MATERIALS_TRIANG2	Composite material name of each **TRIANG2** element	i s
MATERIALS_TRIANG3	Material name of each **TRIANG3** element	i s
MATERIALS_TRIANG4	Material name of each **TRIANG4** element	i s
MATERIALS_TRIANG5	Material name of each **TRIANG5** element	i s
MATERIALS_QUAD1	Material name of each **QUAD1** element	i s
COMP_MATERIALS_QUAD3	Composite material name of each **QUAD3** element	i s
MATERIALS_TETRAH1	Material name of each **TETRAH1** element	i s
MATERIALS_BLOCK1	Material name of each **BLOCK1** element	i s
REMESH_DATA	Defines the beginning of NURBS definition	

Table D.5:c) The mesh definition file keywords and keyword data.

D.1.4 The mesh parameter file

The mesh parameter file (**.mpr**) defines the element mesh parameters of each element according to its element type (Table D.6). The number of mesh parameters for each element type must be equal to the number of elements defined in the mesh definition file. There is a possibility of defining the mesh parameters on a nodal basis. In this case the **NODAL_MESHPARAMS** keyword should be defined and then the mesh parameter for each mesh-point. An example file of element mesh parameters is shown in Table D.7 for Table D.2

D.1.5 The element geometric definition file

The element geometric definition file **.gmf** defines the boundary of a mesh using Non-Uniform Rational B-Splines (NURBS). This file format is discussed in Section C.4.

D.1.6 Miscellaneous file formats

The E_Library format defines four more file formats: material, domain decomposition, finite element errors and stresses. However these file formats are not relevant for the

Keyword	Keyword Data	Data Type
NODAL_MESHPARAMS	Mesh parameters for each node	i r
MESHPARAMS_LINK1	Element mesh parameter of each **LINK1** element	i r
MESHPARAMS_LINK2	Element mesh parameter of each **LINK2** element	i r
MESHPARAMS_LINK3	Element mesh parameter of each **LINK3** element	i r
MESHPARAMS_LINK4	Element mesh parameter of each **LINK4** element	i r
MESHPARAMS_LINK5	Element mesh parameter of each **LINK5** element	i r
MESHPARAMS_TRIANG1	Element mesh parameter of each **TRIANG1** element	i r
MESHPARAMS_TRIANG2	Element mesh parameter of each **TRIANG2** element	i r
MESHPARAMS_TRIANG3	Element mesh parameter of each **TRIANG3** element	i r
MESHPARAMS_TRIANG4	Element mesh parameter of each **TRIANG4** element	i r
MESHPARAMS_TRIANG5	Element mesh parameter of each **TRIANG5** element	i r
MESHPARAMS_QUAD1	Element mesh parameter of each **QUAD1** element	i r
MESHPARAMS_QUAD3	Element mesh parameter of each **QUAD3** element	i r
MESHPARAMS_TETRAH1	Element mesh parameter of each **TETRAH1** element	i r
MESHPARAMS_BLOCK1	Element mesh parameter of each **BLOCK1** element	i r

Table D.6: The element mesh parameters file keywords and keyword data.

```
# An example element mesh parameters file

MESHPARAMS_TRIANG1
  1   5.3456
  2   3.4557
  3   3.4556
  4   0.456
```

Table D.7: Definition of mesh parameters for the example in Table D.2.

mesh generation therefore they are not discussed in this book.

D.2 Viewing and printing E_Library files

To view E_Library files there are two applications:

- **e_plot32** is a native windows 32bit graphical user interface program. It will run on Windows95, Windows98 and Windows NT 4, Windows 2000.
- **eplx** is an X Window program using the Motif library. It should run on all Unix machines, however it was tested on Solaris, IRIX and Linux.

These programs provide a viewer for all sorts of meshes in the E_Library mesh definition format. The user can view the mesh in the following plot configurations:

- plain mesh plot: shows the mesh connectivity;
- subdomain plot: colours the different subdomains according to the subdomain information stored in the appropriate subdomain file;
- stress plot: shows a colour representation of all the stresses that are available in the stress file (only in e_plot32);
- FEerror plot: plots the distributions of the finite element errors as specified in the FEerror file (only in e_plot32);
- mesh parameter plot: makes a plot of the mesh parameters as stored in the mesh parameter file (only in e_plot32);.

The e_plot32 program can also print directly from the menu and also has an export facility to the `*.wmf` format which can be read by many vector graphics packages such as CorelDraw and AutoCAD.

D.2.1 Other limitations of the programs e_plot32

- All related data files (such as subdomains, stresses, etc.) should have the same basename as the input `*.mdf` file with the appropriate extensions.
- For meshes containing `TRIANG2` elements: only the stresses stored for `layer 0` are shown in a stress plot.
- When exporting a subdomain plot to `*.wmf` format the subdomain colours are converted to grayscale. This does not happen with the stress, FEerror or mesh parameter plots.
- An export to `*.wmf` might distort the scaling of the image. Scale the result by a factor 100 to obtain normal sizes. However this depends on the graphics package you use.
- e_plot32 will only accept element mesh parameters for a mesh parameter plot.

D.3 Creating meshes with curves: NURBSVIEW

The **nurbsview** program was developed to handle coarse meshes with curved boundaries. The program is written in Tcl/Tk and therefore it should run on all platforms. The program also requires the **etli** library, which is a thin abstraction layer between E_Library and Tcl/Tk. The program allows the creation and editing of link, triangle and quad elements and most importantly the creation and editing of "NURBS" curves. The resulting mesh can be written out into a mesh file and directly used with the mesh generator program.

Bibliography

[1] Sziveri, J., Seale, C., Editors, *E-lib User's Guide*, SECT Research Group, Heriot-Watt University, Edinburgh, UK, 1997.

Appendix E

Reconnection Patterns

In addition to the reconnection pattern given in Section 6.6.2.2 for the insertion of a node that lies on the side of a triangle, two alternative reconnection patterns are presented here.

If the new node (which is given number n_p) is found on side 1 of triangle t_t, Figure E.1 should be used in conjunction with the pattern from Table E.1. If the new node (which is given number n_p) is found on side 2 of triangle t_t, Figure E.2 should be used in conjunction with the pattern from Table E.2.

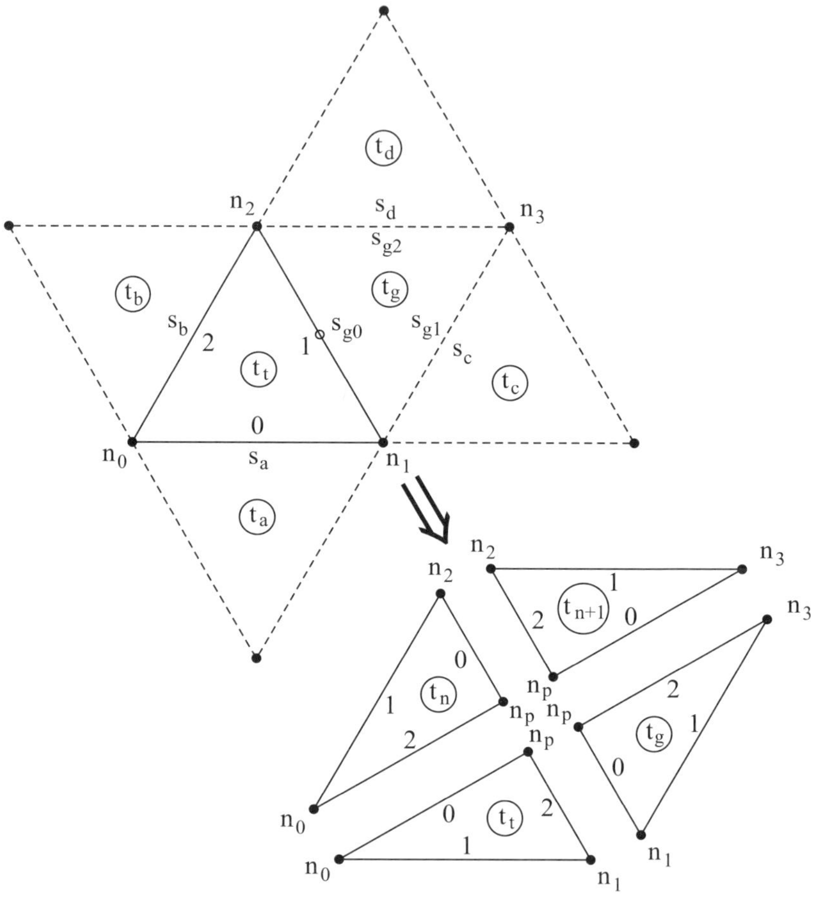

Figure E.1: Reconnecting a node on side 1 of a triangle.

Element matrix before insertion

		0	1	2	3	4	5	6	7	8
$\dots$										
t_t	$\longrightarrow$	n_0	n_1	n_2	t_a	t_g	t_b	s_a	s_{g0}	s_b
t_g	$\longrightarrow$	$\{\ n_3$	n_2	$n_1\ \}_{s_{g0}}$	$\{\ t_d$	t_t	$t_c\ \}_{s_{g0}}$	$\{\ s_d$	1	$s_c\ \}_{s_{g0}}$
$\dots$										
t_a	$\longrightarrow$	$\{\ \star$	n_1	$n_0\ \}_{s_a}$	$\{\ \star$	t_t	$\star\ \}_{s_a}$	$\{\ \star$	0	$\star\ \}_{s_a}$
t_b	$\longrightarrow$	$\{\ \star$	n_0	$n_2\ \}_{s_b}$	$\{\ \star$	t_t	$\star\ \}_{s_b}$	$\{\ \star$	2	$\star\ \}_{s_b}$
t_c	$\longrightarrow$	$\{\ \star$	n_3	$n_1\ \}_{s_c}$	$\{\ \star$	t_g	$\star\ \}_{s_c}$	$\{\ \star$	s_{g1}	$\star\ \}_{s_c}$
t_d	$\longrightarrow$	$\{\ \star$	n_2	$n_3\ \}_{s_d}$	$\{\ \star$	t_g	$\star\ \}_{s_d}$	$\{\ \star$	s_{g2}	$\star\ \}_{s_d}$
$\dots$										

Element matrix after insertion

		0	1	2	3	4	5	6	7	8
$\dots$										
t_t	$\longrightarrow$	n_p	n_0	n_1	t_n	t_a	t_g	2	s_a	0
t_g	$\longrightarrow$	n_p	n_1	n_3	t_t	t_c	t_{n+1}	$\{\ 2$	s_c	0
$\dots$										
t_a	$\longrightarrow$	$\{\ \star$	n_1	$n_0\ \}_{s_a}$	$\{\ \star$	t_t	$\star\ \}_{s_a}$	$\{\ \star$	1	$\star\ \}_{s_a}$
t_b	$\longrightarrow$	$\{\ \star$	n_0	$n_2\ \}_{s_b}$	$\{\ \star$	t_n	$\star\ \}_{s_b}$	$\{\ \star$	1	$\star\ \}_{s_b}$
t_c	$\longrightarrow$	$\{\ \star$	n_3	$n_1\ \}_{s_c}$	$\{\ \star$	t_g	$\star\ \}_{s_c}$	$\{\ \star$	1	$\star\ \}_{s_c}$
t_d	$\longrightarrow$	$\{\ \star$	n_2	$n_3\ \}_{s_d}$	$\{\ \star$	t_{n+1}	$\star\ \}_{s_d}$	$\{\ \star$	1	$\star\ \}_{s_d}$
$\dots$										
t_n	$\longrightarrow$	n_p	n_2	n_0	t_{n+1}	t_b	t_t	2	s_b	0
t_{n+1}	$\longrightarrow$	n_p	n_3	n_2	t_g	t_c	t_n	2	s_d	0
$\dots$										

Table E.1: Reconnecting a node on side 1 of a triangle.

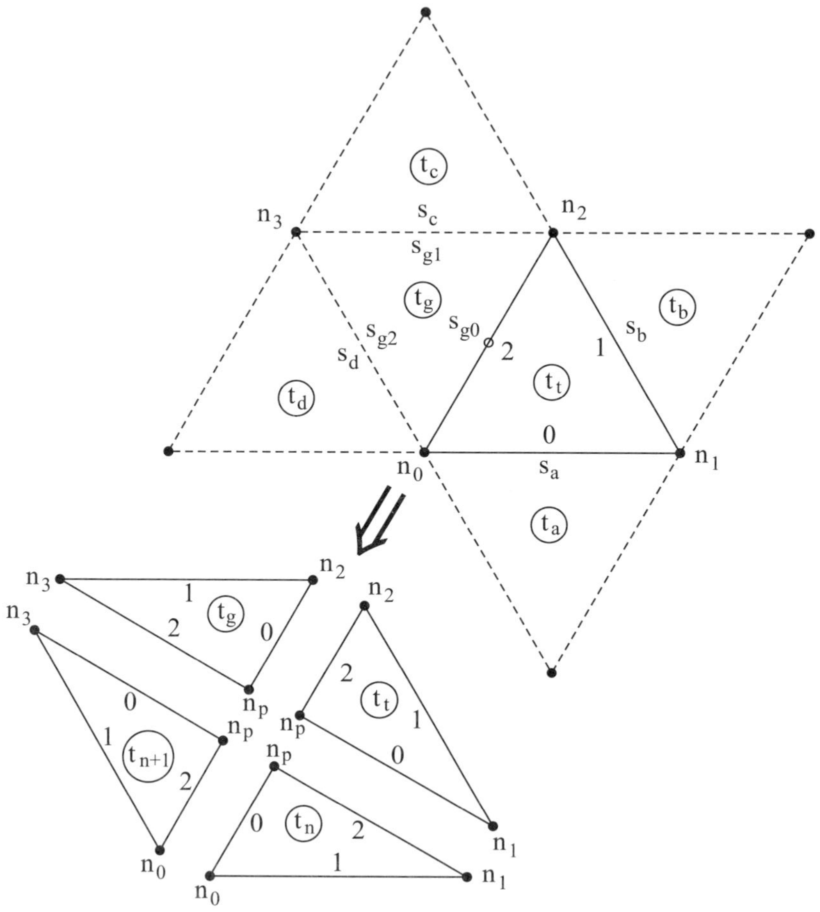

Figure E.2: Reconnecting a node on side 2 of a triangle.

Element matrix before insertion

		0	1	2	3	4	5	6	7	8
$\dots$										
t_t	$\longrightarrow$	n_0	n_1	n_2	t_a	t_b	t_g	s_a	s_b	s_{g0}
t_g	$\longrightarrow$	$\{\ n_3$	n_0	$n_2\ \}_{s_{g0}}$	$\{\ t_d$	t_t	$t_c\ \}_{s_{g0}}$	$\{\ s_d$	2	$s_c\ \}_{s_{g0}}$
$\dots$										
t_a	$\longrightarrow$	$\{\ \star$	n_1	$n_0\ \}_{s_a}$	$\{\ \star$	t_t	$\star\ \}_{s_a}$	$\{\ \star$	0	$\star\ \}_{s_a}$
t_b	$\longrightarrow$	$\{\ \star$	n_2	$n_1\ \}_{s_b}$	$\{\ \star$	t_t	$\star\ \}_{s_b}$	$\{\ \star$	1	$\star\ \}_{s_b}$
t_c	$\longrightarrow$	$\{\ \star$	n_3	$n_2\ \}_{s_c}$	$\{\ \star$	t_g	$\star\ \}_{s_c}$	$\{\ \star$	s_{g1}	$\star\ \}_{s_c}$
t_d	$\longrightarrow$	$\{\ \star$	n_0	$n_3\ \}_{s_d}$	$\{\ \star$	t_g	$\star\ \}_{s_d}$	$\{\ \star$	s_{g2}	$\star\ \}_{s_d}$
$\dots$										

Element matrix after insertion

		0	1	2	3	4	5	6	7	8
$\dots$										
t_t	$\longrightarrow$	n_p	n_1	n_2	t_n	t_b	t_g	2	s_b	0
t_g	$\longrightarrow$	n_p	n_2	n_3	t_t	t_c	t_{n+1}	$\{\ 2$	s_c	0
$\dots$										
t_a	$\longrightarrow$	$\{\ \star$	n_1	$n_0\ \}_{s_a}$	$\{\ \star$	t_n	$\star\ \}_{s_a}$	$\{\ \star$	1	$\star\ \}_{s_a}$
t_b	$\longrightarrow$	$\{\ \star$	n_2	$n_1\ \}_{s_b}$	$\{\ \star$	t_t	$\star\ \}_{s_b}$	$\{\ \star$	1	$\star\ \}_{s_b}$
t_c	$\longrightarrow$	$\{\ \star$	n_3	$n_2\ \}_{s_c}$	$\{\ \star$	t_g	$\star\ \}_{s_c}$	$\{\ \star$	1	$\star\ \}_{s_c}$
t_d	$\longrightarrow$	$\{\ \star$	n_0	$n_3\ \}_{s_d}$	$\{\ \star$	t_{n+1}	$\star\ \}_{s_d}$	$\{\ \star$	1	$\star\ \}_{s_d}$
$\dots$										
t_n	$\longrightarrow$	n_p	n_0	n_1	t_{n+1}	t_a	t_t	2	s_a	0
t_{n+1}	$\longrightarrow$	n_p	n_3	n_0	t_g	t_d	t_n	2	s_d	0
$\dots$										

Table E.2: Reconnecting a node on side 2 of a triangle.

Index

Author Biographies

Barry Topping is Professor of Computational Mechanics at Heriot-Watt University, Edinburgh. His research is concerned with the application of computational technology to structural engineering. The technology includes parallel processing and artificial intelligence techniques. Current application areas include: the design and analysis of offshore steel structures; the analysis of stressed membrane structures; the simulation of gas explosions in buildings; and the analysis of non-linear structural systems.

Barry Topping has been awarded a number of European, UK Research Council, Department of the Environment and Industrial Research Contracts. He is the author or editor of over a hundred technical publications concerned with various aspects of engineering computation and has undertaken computational consultancy for government authorities, consulting engineers and contractors. He has given lectures in Europe, China and the United States of America. In 1990 he was awarded a Telford Premium, by the Institution of Civil Engineers, London, for a paper concerned with structural engineering re-analysis. In 1978 he was appointed Lecturer in Civil Engineering at Edinburgh University and in 1988 Senior Lecturer in Structural Engineering at Heriot-Watt University. From 1986-87 he was Von Humboldt Research Fellow at Stuttgart University, Germany. In 1989, he was appointed Reader at Heriot-Watt University and from 1990-1995 Professor of Structural Engineering. He is: Co-Editor of "Structural Engineering Review" (publishers: Pergamon Press); Co-Editor of "Advances in Engineering Software incorporating Computing Systems in Engineering" (publishers: Elsevier Science Ltd); and a member of the Editorial Board of a number of other scientific journals including: Computers & Structures, Engineering Optimization, Structural Optimization and Engineering Applications of Artificial Intelligence.

Jelle Muylle graduated as a civil engineer at the K.U.Leuven in Belgium. His undergraduate thesis involved structured mesh and cutting generation for tensile membrane structures. He then started a Ph.D. at Barry Topping's Structural Engineering Computational Technology Research Group at Heriot-Watt University, Edinburgh. His research focuses on the parallellisation of contact detection algorithms for the analysis and design of cable and membrane structures.

Currently Jelle works for an engineering consultancy firm, Computational Technology Solutions, where he designs real-time simulation software for underwater cable and pipe-laying applications. He is also involved in the design of cable and membrane structures, and their visualisation.

Peter Iványi was awarded his MSc in Civil Engineering by the Technical University of Budapest, Hungary in 1995. In 1996 he commenced his research into Computational Mechanics with Professor Topping at the Department of Mechanical Engineering of Heriot-Watt University. He was awarded his Ph.D. in 2002. Currently he is a Research Associate and he is responsible for the installation, management and administration of the high performance computing facility at the Department of Mechanical Engineering of Heriot-Watt University. He is the author of a number of papers concerning the design of cable-membrane structures, mesh generation and high performance computing.

Roman Putanowicz was awarded his MSc in Mechanical Engineering by Cracow University of Technology, Poland in 1997. In 1997 he joined the Institute of Computer Methods in Civil Engineering at Cracow University of Technology as a Research Assistant. Between 1999 and 2002 he was on a sabbatical leave at the Structural Engineering Computational Technology Research Group at the Department of Mechanical Engineering of Heriot-Watt University, where he conducted a research under the supervision of Professor Topping. In 2002 he returned to Cracow and currently is working towards his Ph.D. in the area of Problem Solving Environments for Computational Mechanics. He teaches Numerical Methods, Computer Methods and programming.

Biao Cheng was a research student in the Department of Mechanical and Chemical Engineering at Heriot-Watt University. In 2001 he was awarded an MPhil for research undertaken with the Structural Engineering Computational Technology Research Group, on Finite Element Analysis, Mesh Generation and Parallel Computing. He has published papers in international journals and made a number of conference presentations. After graduation he worked for an oil consulting company and is responsible for using numerical and analytical methods for well pressure, temperature analysis, oil field production network optimization, and multi-phase chemical simulation.

SAXE-COBURG
PUBLICATIONS
mmiv